Overcoming Math Anxiety

Other books by Sheila Tobias:

What Kinds of Guns Are They Buying for Your Butter? A Beginner's Guide to Defense, Weaponry, and Military Spending (with Peter Goudinoff, Stefan Leader, and Shelah Leader)

Succeed with Math: Every Student's Guide to Conquering Math Anxiety

Women, Militarism, and War: Essays in History, Politics, and Social Theory (with Jean Bethke Elshtain)

They're not Dumb, They're Different: Stalking the Second Tier

Revitalizing Undergraduate Science: Why Some Things Work and Most Don't

Breaking the Science Barrier (with Carl T. Tomizuka)

Overcoming Math Anxiety

REVISED AND EXPANDED

Sheila Tobias

W.W. Norton & Company New York • London

Printed in the United States of America
First published as a Norton paperback 1995

The text of this book is composed in Cheltenham Light
with the display set in Antique Olive Bold. Composition and
manufacturing by Haddon Craftsmen Inc.
Book design by Charlotte Staub

Library of Congress Cataloging-in-Publication Data

Tobias, Sheila.
Overcoming math anxiety : revised and expanded / Sheila Tobias.
p. cm.
Includes index.
1. Mathematics—Study and teaching—Psychological aspects.
I. Title.
QA11.T67 1994
370.15′651—dc20 93-3648

ISBN 0-393-31307-7
W. W. Norton & Company, Inc., 500 Fifth Avenue, New York, N.Y. 10110
W. W. Norton & Company Ltd., 10 Coptic Street, London WC1A 1PU

3 4 5 6 7 8 9 0

To Paul and Rose Tobias,
Parents, Mentors, and Friends

Contents

Preface to the New Edition

Quite as much as fifteen years ago, when *Overcoming Math Anxiety* was first published, math anxiety remains a political issue. The facts are these: Millions of adults are blocked from professional and personal opportunities because they fear or perform poorly in mathematics. Most of these adults are capable of learning more mathematics. Theirs, as I argue in chapter 2, is not a failure of intellect, but a failure of nerve.

For most people, mathematics is more than a subject. It is a *relationship* between themselves and a discipline purported to be "hard" and reserved only for an elite and powerful few. Thus, all people endure some mathematics anxiety, but it disables the less powerful—that is, women and minorities—more. There are certainly cures for math anxiety. In the short run, these involve changing attitudes and exploding myths about who can do mathematics and how mathematics competence is measured; in the long run, they require changing popular perceptions about mathematics.

Twenty years ago, I began to ask the question: Why do otherwise intelligent adults, people who do well in subjects they like, have a specific disability in mathematics? The question had never before been taken seriously, because it was assumed—wrongly, I believed—that to do mathematics, certainly at the college level, if not before, people like ourselves need to have a special gift for mathematics, a "mathematical mind." The establishment of math-anxiety clinics in colleges and continuing-education programs around the country proved, instead, that most average students have all the cognitive equipment they need to do advanced algebra, intermediate-level statistics, and college calculus. The problem was then, and in many quarters still remains, they just don't *believe* they do.

My task as a researcher-activist was to prove, first to the students my team and I were trying to serve, later to the public at large, that it was anxiety, not lack of "ability," that was getting in their way. Also to demonstrate to the math-teaching community that such people exist and should not be written off. I believed then, and I still believe, that math anxiety is a constructive diagnosis. You can't do anything about being "dumb in math." You can do a great deal about being fearful.

Looking back on the fifteen years that have passed since this book was first conceived and published, I am struck by how much has changed, and yet how much remains the same.

In 1978, the personal computer had not yet been invented; pocket calculators were available, but were not yet built into mathematics education for the young. The math-education community was still insisting that children learn by rote to do the kinds of calculations we all now rely on calculators to do for us. Hundreds of classroom hours were still being spent on tedious memorization and drill—thirty weeks alone on long division. Women were only just beginning to lobby for their right to participate fully in the workplace and to be treated on equal terms with men. More important, women were only just beginning to *believe* they had the capacity and the taste for "nontraditional" jobs and careers in management that in many cases rest on quantitative skills and on a familiarity with statistics and probability.

No one was paying much attention to mathematics achievement or science literacy overall. It was assumed by parents, teachers, and school boards alike that children who enjoyed mathematics would do mathematics, and everyone else didn't have to. Sputnik had shaken our complacency, but that was much earlier. The Soviet Union was still our military rival in 1978, the Japanese threat to our economy just over the horizon. Hence, the link between mathematical competence and economic productivity was not yet understood. Who cared whether the United States as a whole was first in the world in mathematics (as President Bush, in 1991, promised we would be by the year 2000)? Mathematics competence was and would remain a preserve of the few.

If failure to do well at mathematics characterized women and minorities, that was *their* problem. Experts who should have known better blamed the victim, who, it was intoned, just didn't have a *mathematical mind.* Parents believed this, youngsters believed this. And so, long before they had to, millions of people gave up on themselves and on math. Worse yet, a powerful myth still circulated, raised by certain irresponsible researchers to dogma, that there just might be a "male math gene."

I was an administrator of a formerly all-male college in the 1970s. Wesleyan University had recently gone coed and had attracted a wonderfully able and ambitious cohort of women students, whose interests were pretty evenly distributed over the rich and varied curriculum offered by the college. But four years later, when I looked over the transcripts of these women students, I noticed their slippery slide off the quantitative. Would-be economics majors, once they were told that economics required two years of college mathematics, found their way out of economics into other majors. Psychology majors, confronting a junior-year "research-methods" course in probability and data analysis, switched to sociology. Women who had come to college seriously thinking about medicine found the chemistry co-requisite for biology or the premed physics requirement too difficult—or, better said, too *scary,* for their transcripts showed no failing grades in these subjects, just blanks where mathematics or physics or statistics should have been. And below, in the boxes marked "majors," a change of goals.

Something was keeping these young able women (and a sizable minority of young men) from achieving what they had set out to do. Could it be, as some of their instructors believed, that, despite general intelligence and high achievement in other disciplines, certain people are just congenitally *dumb in math?* Or was it *fear of math* that was getting in their way?

Out of these reflections came my determination to test a hypothesis: that anxiety, rather than incompetence, controlled our students' academic choices. I set out to explore this hypothesis first by interviewing students, then by confronting their instructors, and finally by learning what counseling psychology has to teach us about phobias and avoidance behavior more generally. Within a year, a group of us established a Math Clinic at Wesleyan that became a model and a stimulus for some new thinking about math education. When *Ms.* magazine published my article "Math Anxiety: Why Is a Smart Girl like You Counting on Your Fingers?" in 1976, we were ready to tell the world: You don't have to be a genius to do mathematics. All you need is confidence, persistence, a taste for hard work, and *math mental health*—the willingness to learn the math you need when you need it.

Overcoming Math Anxiety has been updated to reflect new findings and some of the positive effects of our early work. Hundreds of studies confirmed my hypothesis that perceived incompetence is often the result of common myths about mathematics: Myth 1, that mathematics ability is inherent; Myth 2, that mathematical insight comes instantly if it comes at all; Myth 3, that only the very few can do mathematics; and Myth 4, that mathematics is a male domain. Gifted young women are no longer hiding their skills in and enjoyment of mathematics and the subjects mathematics serves. At last count, nearly 31 percent of the incoming medical students in the country are women; enrollment among women in undergraduate engineering has risen from virtually none to nearly 13 percent;[1] and individual women are excelling in fields once believed to be "men's work," from piloting military and commercial aircraft to high finance.

Still, according to one longitudinal study of children who score in the top 1 percent in math ability at age eleven, only 8 percent of the males and only 1 percent of the females ever make it to the Ph.D. in science, engineering, or mathematics.[2]

To its credit, the mathematics-education community has responded positively to criticism. A new set of mathematical Standards is making its way through the K–12 mathematics curriculum, the result of a massive infusion of new ideas about mathematics teaching and learning contributed by mathematicians and mathematics educators alike. Gone are the thirty weeks of practice in long division, an absence of real-world applications of arithmetic and algebra, the pretense that the calculator is not here to stay. In their stead (in some but not yet all schools) is a new emphasis on problem solving, mathematical creativity, open-ended questions, and discussion about mathematical *ideas*. The country is paying much more attention to who's in and who's out of the mathematics mainstream. It turned out that not just women and minorities were being disserved by mathematics education. Everyone has a right to be mathematically competent.

This is not to imply that there hasn't been *resistance* to the notion that anyone can learn mathematics. Hardly a year goes by without some "new finding" about sex differences in brain organization that shows males to be advantaged over females,[3] or allegations that one race or cultural group is superior in mathematics to another. One ongoing research group continues to find (and to report to a population made uncomfortable by the idea that women may be equal to men) that, of the ablest in mathematics, as measured by performance on standardized tests, the top males *on average* outperform the top females.[4] (These "averages" don't stop some young women from outperforming *all* young males in the test sample.) It's not just maleness that "predicts" mathematical ability, according to these researchers. Myopia, immune disorders, acne, and left-handedness also show up in the mathematically able.[5] No wonder girls more than boys drop out of these programs, says one observer. With their acne, thick glasses, snively noses, and awkwardness, these boys don't seem much fun to be with.

Nor is math underachievement on the wane. On the contrary. Only about 8 percent of American high-school students elect to study higher mathematics—that is, precalculus and calculus.[6] Although nearly 60 percent are now taking high-school chemistry (a considerable increase over the past dozen years), only 19 percent are in high-school physics—the most mathematical of the elementary sciences. Millions of Americans are still letting fear of mathematics limit their career options, their pleasures in life. And, as the newspaper-reading public knows well, the average American seventeen-year-old is woefully underprepared, in comparison with students elsewhere, in mastery of mathematical concepts.

As long as parents, teachers, athletes, and entertainers publicly indulge in fear or indifference to mathematics, and as long as people who succeed at mathematics claim an innate superiority over people who don't, the myths surrounding mathematics and the math anxiety that is the consequence of these myths will probably not go away.

Of the hundreds of studies that have been done since 1978, the following have been particularly useful to me in revising this volume: On girls and mathematics: Susan Chipman et al., eds., *Women and Mathematics: Balancing the Equation;*[7] and Elizabeth Fennema and Gilah C. Leader, eds., *Mathematics and Gender.*[8] On sex differences in learning styles and the classroom "climate" for girls more generally in school: Mary Field Belenky et al., *Women's Ways of Knowing;*[9] and the American Association of University Women's *How Schools Shortchange Girls.*[10] On mathematics teaching and learning: Lynn Steen, *Everybody Counts: A Report to the Nation on the Future of Mathematics Education;*[11] and *Moving Beyond Myths: Revitalizing Undergraduate Mathematics.*[12] And on the new mathematical Standards for kindergarten through twelfth grade, the publications by the Mathematical Association of America and the National Council of Teachers of Mathematics. On mathematics illiteracy and its consequences among adults of both sexes, John Allen Paulos' *Innumeracy.*[13]

The advent of the computer has been a mixed blessing, as researchers discover that more males than females attend computer camps and have home computers for their personal use; also that children who attend less advantaged schools are taught keyboarding and not programming on their school computers.[14] Jo S. Sanders and Antonia Stone's *The Neuter Computer*[15] and Lynne Alber and Meg Holmberg's *Parents, Kids, and Computers*[16] deal with this issue's impact on young children. C. Dianne Martin and Erich Murchie-Beyma's *In Search of Gender Free Paradigms for Computer Science Education*[17] provides recommendations for college teachers.

I am especially indebted to Sanford Segal, professor of mathematics, University of Rochester, and Persis Herold, director of the Math Learning Center of Washington, D.C., for invaluable suggestions as to the revision of this volume, and for their careful reading of the revised text.

For descriptions of current math-anxiety programs and adult re-entry mathematics, I have either interviewed in depth or monitored a number of ongoing programs, and owe particular thanks to Jean Smith, Wesleyan's first math-anxiety instructor, still offering mathematics "detox" to adult students in Augusta, Maine; Kathryn Brooks and Ashley DuLac of the University of Utah's math-anxiety workshop; Frances Rosamond, who is responsible for all of National University's programs in math-for-adults; Cindy Arem and Debbie Yoklic of Pima County Community College's math-anxiety program; and Gary Horne, Pamela Reavis, and Mary Ellen Hunt, who conducted a math-anxiety course especially for me to observe at the University of Arizona in the fall of 1992. Susan Newcomer and Jacqueline Raphael provided writing and editorial assistance for some of the newer sections; Mary Ellen Hunt and Gary Horne, my principal assistants, provided many ideas and much mathematical insight and acumen.

Edwin Barber, my first and current editor at Norton, and Gloria Stern, my agent, have been with this book since its inception. It was Gloria Stern who first saw that my work on math anxiety at Wesleyan deserved a wider public, and Ed Barber who helped shape my message fifteen years ago and inspired this new edition.

Most of all, I am indebted to the math-anxious victims who have always provided me, freely and without inhibition, their very personal math autobiographies. I only hope that my work serves them as productively as their efforts to overcome their math anxiety have served me.

Tucson, Arizona
August 1993

Notes

1. United States Census, 1986.
2. David Lubinski and Camilla Persson Benbow, "Gender Differences in Abilities and Preferences Among the Gifted: Implications for the Math-Science Pipeline," *Current Directions in Psychological Science,* vol. I, no. 2 (April 1992), p. 64.
3. One of the latest is the claim that sex hormones affect the way women and men solve intellectual problems. See Doreen Kimura, "Sex Differences in the Brain," in a special issue of *Scientific American* devoted to "Mind and Brain," *Scientific American,* Sept. 1992, pp. 119 ff. Also see chap. 3 of this book.
4. Julian C. Stanley and Camilla P. Benbow, "Gender Differences on Eighty-six National Standardized Aptitude and Achievement Tests," paper delivered at the University of Iowa, May 17, 1991.
5. Kimura, "Sex Differences in the Brain," p. 122. See also "Autoimmunity in Left-Handers," *Science,* vol. 217 (July 9, 1982), p. 141.
6. *Trends in Academic Progress—The Nation's Report Card* (Washington, D.C.: National Center for Educational Statistics, January 1992), p. 92.
7. Susan Chipman, L. R. Brush, and D. M. Wilson, eds., *Women and Mathematics: Balancing the Equation* (Hillsdale, N.J.: Lawrence Erlbaum, 1985).

8. Elizabeth Fennema and Gilah C. Leader, eds., *Mathematics and Gender* (New York: Teachers College Press, 1990).
9. Mary Field Belenky, Blythe McVicker Clinchy, Nancy Rule Goldberger, and Jill Mattuck Tarule, *Women's Ways of Knowing* (New York: Basic Books, 1986).
10. American Association of University Women, *How Schools Shortchange Girls* (Wellesley, Mass. Wellesley College Center for Research on Women, 1992).
11. Lynn Steen, ed., *Everybody Counts: A Report to the Nation on the Future of Mathematics Education* (Washington, D.C.: National Academy Press, 1989).
12. Mathematical Sciences Board, *Moving Beyond Myths: Revitalizing Undergraduate Mathematics* (Washington, D.C.: National Academy Press, 1991).
13. John Allen Paulos, *Innumeracy: Mathematical Illiteracy and Its Consequences* (New York: Hill and Wang, 1988).
14. Inner-city schools have fewer computers and make less creative use of them. See Charles Piller and Liza Weiman, "America's Computer Ghetto," *New York Times,* Aug. 7, 1992. Piller and Weiman are editors at *Macworld* magazine.
15. Jo S. Sanders and Antonia Stone, *The Neuter Computer* (New York: Neal Shoman, 1986).
16. Lynne Alber and Meg Holmberg, *Parents, Kids, and Computers* (Berkeley, Calif.: Lawrence Hall of Science, 1984).
17. C. Dianne Martin and Erich Murchie-Beyma, eds., *In Search of Gender Free Paradigms for Computer Science Education* (Eugene, Oregon: International Society for Technology in Education, 1992).

Preface to the First Edition

If anyone had told me four years ago that I would be involved in any way with mathematics, I would have thought the idea absurd. Like many women of my generation, I managed to avoid math and science, and even to avoid thinking about why I had driven these important and interesting fields out of my life. Until I was thirty years old, I never dated a scientist, an engineer, or a math major. My math avoidance extended even to my social life.

Not that I was incompetent. I did well enough in sixth grade, when word problems began to stump my friends. I mastered elementary and intermediate algebra and liked geometry. Math courses were especially attractive to an uncertain teenager because the subject produced right answers that no unfriendly teacher could take away from you. There was a kind of certainty in mathematics that I needed then, and I felt gratified when things came out right.

During high school, things changed. A powerful and subtle pressure went to work on me. For reasons I did not then understand, I began to feel more comfortable not doing math at all. In college, and for years at work, I persuaded myself that I was interested in ideas and in people, and I believed for a long time that measurements only interfered. Being verbal and quick to pick up concepts, I managed to get away with what I finally had to concede was a yawning gap in my understanding of the world.

In the mid-1960s, I became a feminist and subscribed heartily to the notion that preference and better pay for technical work was just one more way to put women down. By the mid-1970s, however, I was troubled by the continuing occupational segregation of women. I then believed that women were being tracked into people-oriented and helping professions because such occupations seemed to them and their colleagues to be appropriate to their female roles. My analysis went something like this:

Traditionally, only three roles are seen as appropriate for women: the mother, the wife, and the decorative or pleasing object. The elementary-school teacher, nurse, and social worker are extensions of the mother role. The assistant-to, secretary, and lab technician are extensions of the wife role. The receptionist, stewardess, and public-relations specialist are the decorative roles. My view seemed to be confirmed when, in the 1960s, stewardesses won the right to keep on working after the age of thirty-five, defeating the airlines' argument that flight attendants must be young and attractive to do their jobs. And when a group of receptionists in a New York corporation threatened a "smile boycott" to demonstrate how important it was for them to be decorative, I felt my analysis was right on target.[1]

Although I realized by then that many of the choices women make are really made for them, I did not see for a very long time that women are predestined to study certain subjects and pursue certain occupations not only because these areas are "feminine" but because girls are socialized not to study math.

Then, in the midst of a busy professional life as a feminist and an educator, I was caught up short one day in 1974 when an unheralded and as yet unpublished piece of information crossed my desk. Lucy Sells, a California sociologist, had done the simplest of surveys. She counted the years of high-school math taken by freshmen males and females entering the University of California at Berkeley in 1972. Of the entering males, 57 percent had taken four years of high-school math, but only 8 percent of the females had had the same amount of math preparation. Without four years of high-school math, students at Berkeley were ineligible for the calculus sequence, unlikely to attempt chemistry or physics, and inadequately prepared for intermediate statistics and economics. Since they could not take the entry-level courses in these fields, 92 percent of the females would be excluded from ten out of twelve colleges at Berkeley and twenty-two out of forty-four majors. Instead, they would be restricted to the "feminine" fields: humanities, guidance and counseling, elementary education, foreign languages, and the fine arts. All other options were foreclosed even before these women arrived at Berkeley.[2]

Something clicked. My previous analysis that the attraction of the motherlike, wifelike, and decorative occupations entirely accounted for job segregation no longer seemed adequate. I began to wonder, instead, about the choices women make early on, and about the significance of avoiding math.

Not many moments in a person's life can be captured in memory. We rarely stop in our tracks and consciously change direction. But I am certain now that when I looked up from Lucy Sells' report I was already contemplating math therapy as a way to get women over math avoidance. The possibilities excited me even more than the problem, for if we could get to the bottom of math avoidance and find a way to demystify mathematics, we might not only change the pattern of occupational segregation but challenge some long-standing notions about sex differences in overall ability. If we could demonstrate, instead, that incompetence in mathematics is learned in conformity to sex-role expectations, then we might be able to design effective compensatory programs for women and girls. And this might affect not only female choice and female performance but also female mental health.

And so, in 1974, I began a campaign against math avoidance. The effort has taken me to other countries to observe the effects of differing expectations on young people, and taken me deeply into mathematics and mathematics teaching. I have examined the myths surrounding mathematics, tried out some intervention techniques in an experimental clinic at my university, conferred with others more familiar than I am with how and why people learn, and looked at the implications of math avoidance on American education and values.

This book contains the results of my inquiry. It is not principally a how-to book, except perhaps for chapters 6 and 7, which demonstrate how mathematical ideas become more comprehensible when presented playfully and in words, not symbols. The book is mainly a discussion of how intimidation, myth, misunderstanding, and missed opportunities have affected a large proportion of the population. My principal purpose in writing this book is to convince women and men that their fear of mathematics is the result and not the cause of their negative experiences with mathematics, and to encourage them to give themselves one more chance.

Much of what I have to say stems from introspection, for I am a victim as well as a reporter of math anxiety. Much is drawn from many hours spent with learners and practitioners of mathematics, trying to find out how they feel when they do mathematics. Feelings are, I believe, at the heart of the problem, although we are supposed to leave feelings outside the math classroom. I have also considered the debates surrounding mathematics learning and dealt with the issues of spatial visualization and cognitive styles. Many of my own insights and my overall conviction that math anxiety is curable come from three years observing a math clinic at work. I have seen undergraduates and adults, at first too fearful of math to elect a single quantitative course, to change jobs, or to ask for a promotion, alter their goals and, more important, their self-image as a result of some intervention. I have watched people learn to trust their own intuition again and to recapture the unfettered curiosity of their youth.

Four years ago, when I began, I hypothesized that mathematics anxiety and mathematics avoidance were feminist issues. Now I am not so sure. Observing men has shown me that some men as well as the majority of women have been denied the pleasures and the power that competence in math and science can provide.* The feminists sounded the alarm. But, as a result, people of both sexes are beginning to reassess their mathematical potential. It is to give these people courage and direction that I have written this book.

January, 1978

Notes

1. Sheila Tobias, "Occupational Segregation: The Next Great Hurdle," unpublished, 1973; and Louise Kapp Howe, *Pink Collar Workers*, (New York: G. P. Putnam's Sons, 1977), *passim*.
2. Lucy Sells' article was finally published in 1978 under the title, "Mathematics—A Critical Filter," *Science Teacher*, vol. 45 no. 2 (Feb. 1978). Sells' findings are now disputed; but the shock they initially provided put mathematics on the feminist agenda.

*In this book, meant for people of both sexes, the pronouns "he," "she," "him," and "her" will be used interchangeably unless a particular point about males or females is being made.

Acknowledgments

Some will say that it is too early to write about math anxiety, since the experts have only just begun to examine the nature and causes of math avoidance. I recognize that I will not have the last word on the subject. Nor can I take credit for having the first word. Long before I began to think about mathematics and mathematics avoidance, a number of researchers in education and mathematics had been studying underachievement among women and men in math. I could not have written this book at all without the previous work of Elizabeth Fennema, Julia Sherman, Lucy Sells, Lynn Fox, John Ernest, Mitchell Lazarus, Robert Davis, Peter Hilton, Lenore Blum, Alice Schafer and her colleagues at Wellesley College, Robert Rosenbaum and Lorelei Brush from Wesleyan University, and Stanley Kogelman, Michael Nelson, Dora Helen Skypek, Judith Jacobs, and others. Their published work is cited in the notes following each chapter.

This listing of my predecessors does not imply that they agree with what I say. Some may find my evidence scanty and my speculations uncalled for. Since my methods are humanistic, I am willing to speculate on the basis of what social scientists would consider inadequate evidence. That is because I am as interested in what is possibly true as in what is necessarily true, and as concerned about the symptoms of math anxiety as about causes and correlations. I believe that only with careful scrutiny of the symptoms will we finally understand what it *feels* like to be poor at math.

I expect that my analysis of feelings may contain insights worth investigating with more care, and I hope that the experts will acknowledge the value of my speculation, just as I appreciate their greater caution.

Without support from the Fund for the Improvement of Post-Secondary Education (FIPSE) for the Math Clinic at Wesleyan, and from the Sloan Foundation, which helped support the clinic in 1977–78, there would have been no project from which I could learn. A Ford Foundation grant gave me time off to work on problems of women in higher education in 1976–77, and the German Marshall Fund of the United States paid my expenses while I was observing math anxiety and avoidance in women in four European countries in the summer of 1976. Program officers are not sufficiently appreciated. People like Alison Bernstein of FIPSE, Mariam Chamberlain of the Ford Foundation, Arthur Singer of the Sloan Foundation, and Gerald Livingston of the German Marshall Fund take risks when they support interesting but unproven new ideas, which was all we had to offer them when we began to tackle math anxiety.

In the course of the writing of this book, family and friends have become colleagues, and colleagues friends. I am most deeply indebted to Carlos Stern, who taught me to appreciate the power of mathematics without making me feel dumb. Writers do not usually cite conversations as sources of their material, but it was in countless conversations with him that I began to confront my own anxiety and avoidance of mathematics. Lucy Knight, an education journalist, shares my deep commitment to demystifying all disciplines; from this perspective, she commented upon the book throughout its writing. Persis Joan Herold gave me insight into how children learn, and as a bonus sketched the drawings and charts. Mathematicians and mathematics educators Joel Schneider, Knowles Dougherty, Fred Fischer, and Susan Auslander patiently tried to straighten out my mathematics, and though they are not responsible for my errors, they get credit for what I finally do understand.

Bonnie Donady and Jean Smith of Wesleyan University, Jane S. Stein of Duke University, Joseph Warren and Stanley Kogelman of Mind over Math in New York, Deborah Hughes-Hallett of Harvard, and Charles Stuth and Beverly Prosser Gelwick of Stephens College in Missouri generously shared their experiences in running math courses and clinics. My father, Paul J. Tobias, taught me to think creatively about the word problems in chapter 5, invented the Letter Division puzzles also in chapter 5, and in his evaluation of my work continued the shaping of my mind that he began so many years ago.

There is no way to thank the hundreds of people who wrote to me after my article on math anxiety was published in *Ms.* magazine in September 1976, but I drew heavily upon their math autobiographies; nor can I acknowledge all the people who regaled me with quarter-hour-long recollections about learning math when they met me somewhere and found out what I was doing. Some people can be thanked, however. Lorraine Karatkewicz, my administrative assistant and dear friend, made it possible for me to think while managing a clinic and fulfilling my obligations as associate provost. Colin Campbell, president of Wesleyan, and an extraordinary Board of Trustees supported the project unstintingly from the beginning.

Last, I want to thank Robert Rosenbaum, professor of mathematics at Wesleyan and codirector of the Math Clinic, who has taught me more about learning than I can possible acknowledge here and who, at a critical moment in the planning of this project, said "Yes."

Middletown, Connecticut
January 1978

Overcoming Math Anxiety

1 The Primacy of Mathematics, or If I Could Do Math I Would . . .

One day soon after we began working with people who avoid mathematics, a visiting Russian mathematician stopped in at the Math Clinic. It took a while to explain to him what the clinic had been designed to do, how people who have long ago given up on their ability to learn math need to be recaptured and motivated to try again. Finally, he seemed to understand. Then he began to laugh and with a big, generous smile he commented: "You Americans are all the same. You think everybody has to know everything."

In a way this is true, although he needn't have been quite so smug.* The United States graduates more young people from

*Russian children seem to have just as much trouble learning elementary arithmetic as American children. A collection of essays on teaching arithmetic in the

high school than most other nations, and we have a larger proportion of people enrolled in some sort of post-secondary schooling than any other nation in the world. On paper at least, we are committed to the belief that everyone should receive a general education and that specialized training should be postponed until later. Still, not everyone learns everything. Many of us elect to stay within a fairly narrow comfort zone of knowledge. If we do well at reading long nonfiction books and writing essays about them, we take more courses in history. If we excel at learning one foreign language, we take another one or two. If we learn best by handling things, we opt for lab courses. And so on.

By the time we are adults, we are pretty smart in our special fields but feel rather dumb when we move outside our comfort zone. What began as a set of preferences becomes in time a mental prison that makes us feel conflicted and even anxious about stepping outside. It is difficult enough for a young person to be a beginner, dependent on other people to show the way. To be a beginner as an adult takes a special kind of courage and enthusiasm for the subject. That is why I believe that a slight discomfort with mathematics acquired in elementary or secondary school can develop into full-fledged syndrome of anxiety and avoidance by the time one has graduated from school and gone to work.

How do these preferences and antagonisms develop? Are they inevitable? Do we hate math, as so many of us say we do, because one elementary-school teacher made us afraid of it? Did we simply start off on the wrong foot and get stuck there? Or have we accepted someone's assumption about our kind of mind and what we thought was the nature of mathematics? Were we too young, from a neurological point of view, when we were first exposed to numbers? Or was too much going on during our adolescent years, when we should have been concentrating on alge-

---

former Soviet Union reports one child's strategy for getting the right answer as follows: "I add, subtract, multiply, and then divide until I get the answer that is in the back of the book."

bra? Does the fault lie in the curriculum, in the subject, in ourselves, or in the society at large?

There are no easy answers to such questions. But some interesting insights have come from people who have avoided mathematics all their lives. In dozens of "math autobiographies," the nature of math anxiety, its causes, and its consequences begin to emerge.

"If I could do math, I would . . ."*

Ask anyone who "hates math" to complete this sentence and the first response will be incredulity. "What, me do math? Impossible." If prodded, however, people reveal all kinds of unexplored needs. "If I could do math, I would . . . fix my own car." "If I could do math, I would . . . fly an airplane." "If I could do math, I would work the F stops on my camera." "If I could do math . . ."

People who don't know what math is don't know what math isn't. Therefore, fear of math may lead them to avoid all manner of data and to feel uncomfortable working with things. Any mathematician will tell you that you don't need mathematics to work the F stops on the camera, or to fix the car, or even to start your own business.† But fearing math makes us wary of activities that *may* involve ratios (F stops), or mechanics, or neat rows of numbers (bookkeeping). What could we do if we could do math? It depends, of course, on how much math we know and how confident we are of our knowledge. But it is safe to say that if we were comfortable with numbers we could do something we are not doing now.

Most people leave school as failures at math, or at least feeling like failures. Some students are not even given a chance to fail. Identified early as non–college material, they are steered away from precollege mathematics and tracked into business or "general math." But high-school students who are not going on to

*"If I could do math, I would," is a phrase invented by Bonnie Donady, counselor of the Wesleyan Math Clinic.

†F stops are a measure of the lens opening: The higher the number, the smaller the opening.

college need algebra and geometry just as much as the college-bound. Algebra and geometry are not just precollege courses.

People who study the link between mathematics and vocational opportunities believe that knowledge of algebra and geometry divides the unskilled and clerical jobs from the better-paying, upwardly mobile positions available to high-school graduates. Mastery of high-school algebra alone will make the difference between a low score and a high score on most standardized entry-level tests for civil service, federal service, industry, and the armed services. And a higher score, even when jobs are scarce, means a higher entry job and the chance to be earmarked for on-the-job training. Frances Rosamond of National University, among others, estimates that starting salaries go up $2,000 per year for *every mathematics course taken after the ninth grade.*[1]

The terminal high-school graduate is not the only one to suffer vocationally for lack of sufficient math. Just as poignant is the plight of the executive who complains that she cannot get any facts from the figures she encounters. Indeed, if she cannot interpret quantitative data in the form of balance statements, budgets, and records of sales, she is in trouble. As mathematician Lynn Osen puts it, business today needs people "who can understand a simple formula, read a graph and interpret a statement about probability."[2]

It comes as no surprise that fields like engineering, computer programming, accounting, geology, and other technical work all require some mathematics. It is not widely known, however, that mathematics, disguised as "quantitative analysis" or "data handling" or "planning," will be used even in nontechnical work areas. It can be painful to discover in mid-career, like the executive cited above, that we cannot get facts from our figures or that an occupation that looked comfortably free of mathematics at the lower levels will require familiarity with quantitative methods for advancement. Yet those who supervise, plan, and manage in fields such as social work, librarianship, retail sales, school administration, and even publishing now need to be familiar with or at least willing to learn more math.

Thus, even carefully prepared guidance material may be misleading. The information still distributed—particularly to middle-school youngsters, who will soon make decisions as to what mathematics courses to take—tends to list only careers that explicitly require mathematics competency *at the very outset.* Often unmentioned is the "new truth" as expressed by W. W. Sawyer: "The ability to think mathematically will have to become something taken for granted as much as the ability to read a newspaper is at present. Such a change will seem fantastic to some people. But so would universal literacy have seemed absurd a few centuries ago."

The *Dictionary of Occupational Titles,* published regularly by the U.S. Department of Labor, codes occupations by one of six mathematics-competency levels, from arithmetic (levels 1 and 2) through intermediate algebra and geometry (level 3) to algebra, calculus, and statistics (level 5) and higher competence at these subjects (level 6). A quick glance at any page in that volume indicates that, if we restrict ourselves to levels 1 and 2 in mathematics, whole families of occupations will be closed to us.[3]

Another measure of math avoidance can be inferred from vocational interest tests. Typically, these group an individual's personal traits into categories which in turn can be translated into jobs. One of the tests, the Self-Directed Search Inventory, lists as interest types: "realistic," "investigative," "artistic," "enterprising," "social," and "conventional." John Holland, designer of this inventory, also publishes an *Occupation Finder* in which occupations are coded according to the same categories. Engineering, for example, is defined as "realistic-investigative," management as "realistic-enterprising-social," and so on. In fact, far more occupations require realistic, enterprising, and investigative skills than require primarily "social" attributes.[4]

Of course, the characteristics of each occupation are derived from people currently at work. Thus, there is a tautology: if engineering is dominated by "realistic-investigative" types, people in the field will of course report that this configuration of skills and temperament is highly valued in engineering and essential to it. And there may be hidden biases against women and "feminine" traits. John Holland is himself mindful of this and tries to factor out such bias. But when an entire society associates math, science, things, and data with male pursuits, it is difficult to eliminate the bias altogether. And nothing changes the fact that math avoidance is extremely limiting for people at all levels of work. Competence in math is, as sociologist Lucy Sells puts it, truly a vocational filter.[5]

Not everyone who avoided mathematics in school remains incompetent, however. In nearly twenty years of interviewing adults who are math-anxious and those who are not, I have found that some people learn to cope with numbers in interesting and sometimes extraordinary ways. They know they have gaps in their knowledge, but they are sufficiently confident and familiar with the way their own minds work to reconstruct problems and solve them.

In a session for a mixed group of adults, Knowles Dougherty, a specialist in teaching the math-disabled, demonstrated this phenomenon. He began by asking the group to do a simple subtraction problem in their heads:

> A woman is 38 years old. It is now 1993.
> In what year was she born?

"Don't tell me the answer," he said. "But as I go around the room, do tell me how you worked out the solution." The methods were astonishingly varied. One adult male said, "I had that certain feeling I had to get to the nearest 10. So I added 2 to 38 to get 40 and then subtracted 40 from 1993 and then added 2 to the answer." Another person reported that she had adjusted the problem to her own age. Starting with her own year of birth, she added and subtracted until she got 1955 (the correct answer).

Each adult was a little ashamed of his or her system. Everyone assumed that, just as there was only one right answer to math problems, there was probably only one right way to subtract. But Dougherty reassured them that the systems they were using were legitimate algorithms (an algorithm being just a system for getting an answer). If you have to get to the nearest 10, then get there. If you have to use some personal reference point, use it. The fact that most people had not used the method they had been taught in school indicated that they had probably learned a fair amount of mathematics on their own since second grade.

Many adults know all this intuitively and develop ways to work out number problems that make sense to them. In Washington, D.C., a professional woman receives bimonthly account records from her bookkeeper. She can make no sense at all of what they show until she goes off by herself and reorganizes the data. She turns columns into rows and rows into boxes, adjusting the numerical information to fit her thinking style. Then she returns to the discussion with her bookkeeper, facts in hand. Some people need to draw pictures. Some have to speak the numbers or the problem out loud.

These people share a willingness to restructure a problem so that it makes sense to them. The executive who could not get the facts from her figures had not yet learned how to do that.

Why can't everyone make such a healthy accommodation? Why can't we just go out and learn the math we need when we need it? The rest of this book will deal in depth with what people remember about studying math, and why they believe they cannot learn more. Apart from the psychological blocks that may develop over time, some of the issues are these: Math is difficult because it is rigorous and complex. As we advance in math, the notation becomes abstract and general. This adds to its mysteriousness. Besides, there are conflicts between the common everyday use of words and the use of these words in math. Moreover, unless we are concurrently studying science or engineering, math is not as integrated into the rest of the curriculum as are reading, writing, and spelling. Hence, we do not immediately apply the math we learn.

Math typically is not learned as part of a group effort. Since we work by ourselves, the process by which we either get or do not get an insight is obscure. And since we do not learn how our own minds work, we never understand why one idea is difficult for us and the next one not.

Finally, people remember math as being taught in an atmosphere of tension created by the emphasis on right answers, and especially by the demands of timed tests. Their math teachers, who may well have been patient and sincere to begin with, became frustrated by the challenge of getting students to understand "simple" ideas; even if they did not start out this way, many of them eventually became cantankerous, short of patience, and contemptuous of error. Math-anxious adults can recall with appalling accuracy the exact wording of a trick question or the day they had to stand at the blackboard alone, even if these events took place thirty years before.

It will surprise no one that millions do not learn mathematics successfully, however well or poorly it is taught. But help may be on the way.

The most recent attempt to improve the teaching of elementary mathematics is the Standards, a remarkable effort on the part of mathematicians and mathematics educators—eight years in the making—to revise the curriculum radically in the direction of more useful and usable mathematics. Published in 1989 by the Mathematics Association of America and the National Council of Teachers of Mathematics, the Standards is only now making its slow way to school-board adoption and teacher acceptance. With sixteen thousand school districts, each jealously guarding local control, it may be a long time before every child is introduced to the *new* New Math. But when that time comes, much of the trauma and mindless drill that characterized math teaching during the period in which we were growing up could be eliminated.

The new Standards calls for: the teaching of how to think for ourselves, group work at all levels of math education, efficient use of technology, the teaching of paper-and-pencil estimation, more statistics and probability in the early grades, less computational drill and practice, the use of concrete materials in teaching, and more realistic problems.

Imagine that, instead of rote memorization of arithmetic facts, our math teachers had conveyed to us the most important truth: that mathematics is first and foremost a *language,* and that the point of all the stuff we had to learn was to provide us with a way of organizing information so that we could make decisions in our lives. What if a mathematical "proof" in geometry had been presented not as something nerdy people do to prove the obvious but, rather, as a system for reasoning to conclusions using certain tried-and-true linguistic tools and ordinary logic. "The average student and the average adult reason to conclusions all the time," our teacher might have added, letting us know that what we were learning was not altogether *new* and specialized but, rather, a *systems approach* to thinking about all kinds of issues. Mathematics, we would have come to realize, makes that reasoning more precise (and more universal); since there is less room for ambiguity in mathematics, there is less chance for error. Had mathematics been presented in this way, I think, those of us who love precision of language and clear thinking in other fields would not have been allowed to consider our excellence at language arts *irrelevant* to our doing well at math, and we might have been full of enthusiasm for the subject instead of dread.

The Standards promises to present mathematics in just this way: as a thinking and decision-making tool. It directs teachers to start with *concrete* materials, instead of meaningless abstractions, to get children to see that mathematics makes sense of their everyday experience. Besides, as many mathematics educators have known for a long time, having children use concrete materials makes pedagogical sense. "When a child holds three red blocks in one hand and five red blocks in the other, he can *feel* eightness," says mathematics educator Jean Smith, co-author of a new textbook for math-anxious adults.[6] "But there is no reason that same child will *make that connection* when the symbols 5 and 3 are flashed on cards."

It was not only the *content* but the *classroom climate* that made many of us anxious about math. Recognizing that competition increases tension ("Raise your hands when you know the answer, children," or "You'll stay at the blackboard until you find your error"), the Standards encourages teachers to have children work in groups, to explore different approaches to solving problems, and to eschew timed tests and that constant drive to find the one right answer. "The only thing I ever really learned in arithmetic," one of our Wesleyan students told us during a math-autobiography session, "was subtraction and short division. During timed tests, I would be forever watching the clock, subtracting the minutes left, and dividing by the number of questions I still had to answer. When I realized I was never going to finish, I would just put down my pencil and give up." "Going to the blackboard in math class," said another student, who was determined never to take another math class, "was the kind of thing dumb animals knew enough to avoid: turning your back to the enemy while you tried to figure out what you were doing wrong." If the Standards bearers have their way, this kind of classroom practice will never happen again.

The Standards promises to reward thinking and originality in place of mindless memorization. And so it must. A story circulated some years ago that the only mathematics truths tenth-graders were really sure about were these: "If you see a decimal point, move it; if you see a fraction, invert it; if you see a sign, change it." "What I learned in ten years of school mathematics," another of our victims recalled, "was what to do when I *remembered* what to do. What I really needed to know was what to do when I forgot."

Real mathematics—the kind we need for everyday problem solving—involves estimation, at least for starters, so we can anticipate what the solution ought to look like before we punch numbers into our calculators. It should be a personal odyssey, fitting our thinking style, our common sense. When I am stuck, I tend to turn big numbers in problems into smaller ones, problems with much complexity into bite-sized pieces I can chew on one at a time. This gives me something sure to stand on while I am figuring out the rest.

When this book was first published, the cartoonist Charles Schulz, some of whose cartoons I had used in the first edition, flattered me by dedicating four consecutive cartoons to "math anxiety," sprinkled with problems for Peppermint Pattie to solve (see figure 1-1).[7] One of the problems Schulz chose to illustrate math anxiety was this one:

Figure 1-1

We all laughed, of course. Trouble was, I thought after the first flush of pleasure at becoming part of Snoopy's world, I just might be called on to answer that question!

My first grab at the problem triggered the word "factorial" (9 factorial, written 9!, is the product of 9 × 8 × 7 × 6, etc.). But when I did the calculation, the answer was—it seemed to me—much too large (362,880).* Rather than let my uncertainty get the best of me, I tried my system out with a smaller number of books. How many ways might 4 books be arranged? Six books? Slowly, I realized the answers corresponded to the factorials of 4 or 6. I checked my multiplication and waited for some snappy journalist to call me for the answer. The call never came, but I realized, yet again, how important it is to use whatever works for us personally in solving math problems.

If the Standards is ever implemented, tomorrow's children will have lots of practice in calculating probabilities. They will be familiar with permutations and combinations, including the kind of factorial problem I wrestled with above. They will get to word problems in the early grades—and more authentic ones, at that. Gone will be the automatic responses we as children learned by heart—that "left" means to subtract, for example, whatever the context.

> Johnnie's mother gives him $6 to take to the circus. He buys 1 ride for $1.25, 1 ride for $2.75, and 2 rides for 50 cents each. How much money does Johnnie have *left* when he returns home?

*That number is correct if you are obliged to use all 9 books in every arrangement, but if you are allowed to use fewer than 9, the possibilities are many, many more—and the problem more difficult to solve.

After a day at the circus, as mathematician Peter Hilton likes to tell math teachers when he comments on the way math problems appear to their pupils, no kid in the real world would *add* the price of all the rides he took and then *subtract* that sum from what he started with. If any real-life mother or dad were to ask, in the stilted way math-book adults talk, "Welcome back, Johnnie. How much money do you have *left?,*" real-life Johnnies and Jills and Consuelas would surely put their hands in their pockets and *add* (not subtract) the remaining coins. (If they had any, and why should they after a day at the circus?) No wonder those of us who endured the old math left school convinced that, whatever else it might be, math was surely not going to be very useful in the lives we intended to lead.

Why should adults want to relearn arithmetic? Why should those of us who have successfully avoided mathematics all these years bother about it at all? Mathematicians have usually been admired for their intelligence but also somewhat scorned for their otherworldliness. Why is their subject suddenly so central to modern life?

To some extent, mathematics has always played a role in our working lives, particularly at the higher occupational levels. The news is that math is increasingly pervasive and also that these occupational levels are beginning to be reached by different people. No longer are academically bright males from advantaged homes the only ones who are expected to go as far in math as they can. In the last twenty years, females and males from less advantaged backgrounds have been encouraged to take school seriously and to expect advancement on the job. These groups have traditionally had little success with higher math and have "mercifully" been allowed to drop it. Much to their surprise and dismay, as their careers advance, math returns to haunt them.

The primacy of mathematics in today's world, then, exhibits the traditional themes of American history: ever-increasing dependence on technology and ever-increasing democratization of jobs.

As in every other aspect of American life, however, World War II accelerated changes that were already under way in the running of business and government. A typical story, probably neither entirely true nor entirely false, recounts that when Robert McNamara returned from his wartime assignment in logistics and supplies, he sold himself and his entire team to the Ford Motor Company with a promise to bring "systems analysis" to management. He and his colleagues had developed computer systems to keep track of the whereabouts of war materiel. McNamara persuaded Ford that systems analysis, then a new idea, could give management the same kind of control over production, marketing, and distribution. Previously, except for scientific and engineering industries, the only people who generally used mathematics at work were those in investment and finance or accounting and budgeting. Decisions elsewhere were made either by extrapolating from laboriously hand-processed data or by what is now recalled as "flying by the seat of one's pants." Although some amount of intuition always comes into decision-making, the mix of data and intuition was significantly altered by systems analysis.

The benefit of systems analysis, now enhanced by linear programming, is that management has at its fingertips both a precise picture of what is actually going on and a means of *predicting* what would be the result of any change in price or market share (or environmental regulation, for that matter) on output and on profit. By "plugging in" possible changes—both those within and those outside their control—management can make the best, or what is called the "optimal," business decision *in advance* of

Figure 1-2

having actually to experiment with a variety of business strategies, some of which might have negative results.

Once upon a time, a company might use experience or intuition to determine how much of any particular product to manufacture and what price to charge. Today, the question is posed more mathematically: With so many square yards of tin and so many orders for big cans, medium-sized cans, and small cans, each at a certain price and each providing a certain percentage of profit, what *combination* of large, medium-sized, and small cans should we manufacture in a month, a week, or a day to maximize profit? And the answer will be the point (or points) on a graph where three or more mathematical curves intersect. Assisted by the computer, which handles such calculations quickly and without error, business has entered a new era of decision-making. Decisions about when to buy and how much, when to sell and at what price, when to unload because cost of inventory begins to overcome traditional mark-up, when to expand, when to contract, and even when to go out of business are all easier to make with mathematics than without.

Nor can the math avoider escape challenges like these by fleeing to the nonprofit world. Just as applied mathematics has revolutionized business practices, the assessment of nonprofit, socially useful programs has become more quantitative, too.

Before World War II, a technique called benefit-cost analysis was hardly known outside the water-resources area of the federal government. To get some sense of the usefulness of water projects for flood control, irrigation, recreation, and production of electric power, the Congress required that large projects be evaluated in terms of their benefits in relation to their costs. Ideally, the benefits of a dam, for example, should always exceed its costs by some amount.

There are, of course, difficulties in comparing such benefits as irrigation, wilderness recreation, and hydroelectric power, and these conflicting values gave rise to serious debates for years. But benefit-cost analysis remained attractive because it provided one way to compare high-cost projects. Since the Congress usually voted these projects in an omnibus bill, including many other projects that could not be debated individually, the benefit-cost ratio might be the only feature of a project that would be noted when the time came to vote.

Just as benefit-cost analysis was developed to meet a problem in the water-resources area, so was it borrowed to meet similar problems in other areas as the federal government began spending large sums for other kinds of domestic projects.

Today, benefit-cost and its first cousin "cost-effectiveness" are basic tools of program evaluation. They are used in assessing environmental regulation, welfare reform, health measures, and even educational programs. However difficult it is to quantify the value of a bird or the good reputation of an institution, administrators today want to know at least whether a proposed project will cost more than it benefits, even if they decide to ignore such figures for the sake of some larger (or smaller) goal.

Since even in the nonprofit sector projects are always competing for funds, it is important to know their cost-effectiveness: Which project will get more impact, dollar for dollar? Will a dollar spent on infant nutrition have a greater payoff in terms of poor children's intellectual performance than a dollar spent on a Head Start program in an urban school? Will a dollar spent in oil exploration produce more benefits than a dollar spent in home insulation? Measuring impact and comparing the cost-effectiveness of programs has become as necessary a part of social work and prison reform as assessing the costs and benefits of automation has become a part of running a large library.

But even more has changed than the worlds of business and nonprofit enterprises. At most colleges and universities, half or more of the undergraduate majors require the calculus sequence or a course in intermediate statistics. This reflects a change in the nature of research as many of the social sciences embrace techniques for taking large surveys, for handling the masses of data that result from these surveys, and for relating the phenomena these data reveal. Regression analysis, perhaps the most useful of all the statistical techniques being applied to social issues today, allows us to compare answers to several questions in a large, many-faceted study, to find out whether the answers are related in any way—whether age, for example, correlates with a certain opinion on abortion rights; or whether religion, sex, and geographical location are more powerful correlates than age; whether smoking correlates with the incidence of throat cancer; whether College Board scores predict college performance.

In a study of energy conservation, for example, two factors are obviously related to the amount of fuel a family consumes: the weather and the size of the apartment or house. But other factors, such as the price of fuel, whether the renter or the landlord pays for the fuel, whether there is thermostatic control in every room, are also significant. How is one to figure out which of these is most important and which are interdependent?

Since many problems, like this one, involve more than two factors, and since "multiple" regression analysis beyond three factors is virtually impossible without a computer, the common use of these statistical techniques has been growing with the availability of computers and the refinement of computer programming. The computer can do what no human brain could do in a reasonable amount of time: sort and resort large amounts of data, seeking patterns and connections among the elements.

This emphasis on statistical techniques should not of course diminish the importance of human judgment. On the contrary, such techniques make human judgment all the more critical, since wrong questions will produce useless if not dangerously misleading answers. As the computer specialists put it, "Garbage in, garbage out." Still, frequent use of all such techniques in modern research means that the math avoider will be excluded from these career areas. One graduate-school admissions dean put it bluntly when she said, "I'd rather have a math major enter psychology at the graduate level than take in a psychology major who had studied no math."

Most sophisticated and powerful of all the new techniques are the large mathematical models that reproduce mathematically a system, a process, or an entire institution, incorporating information so complex that nobody could possibly retain it all in any number of files. These models are designed to simulate the real world, so that the researcher or manager can try out some idea, such as a shift in price per item or in temperature per beaker, and "see" what effect the change will have on the entire process.

Such a model, for example, was developed for the U.S. Department of Transportation. It was designed to evaluate AMTRAK's punctuality and suggest ways to minimize lateness. The mathematician in charge of the model collected all the arrival and departure information for all stations in a given period, along with the number of passengers who get on and off in each place. His job was to find a way to ascertain not simply how many AMTRAK trains were how late (this would be simple), but how many people were inconvenienced by how much by AMTRAK's lateness. The mathematician's first job was to create a basic unit for measuring passenger inconvenience, such as a unit of passenger-late-minutes-per-mile. This whole process may sound exotic, but if AMTRAK has only enough surplus money one year to fix so many miles of track or to add only so many extra ticket agents or to reschedule only so many stops, the question for management must be "What combination of these efforts will minimize total passenger inconvenience and maximize passenger satisfaction?" Information on the passenger-late-minutes-per-mile for any one segment of track will help.

The same kind of analysis is used by local transportation authorities to determine the placement and timing of traffic lights.

Other, perhaps more ghoulish, but no less important calculations come from the new science of "risk assessment." In calculating costs of "avertible deaths" per dollars spent, the Harvard School of Public Health has been able to do comparisons that reveal how much (or how little) some public policy actually costs in terms of dollars spent per year of life saved—in other words, what it costs our government to avert citizen deaths from particular threats. On this measure, the money we spend to save a year of life by means of vaccinating children against measles is less than $1.00; the cost of a year of life saved by means of coronary-artery bypass surgery is $67,579; and by means of routine radiation control at nuclear power plants, $164,875,379.[8] The enormous range of expenditures shows how powerful our *perception* of danger can be in determining government outlays, and how a campaign to change those perceptions—e.g., to educate people as to the reality of risks—could save a great deal of tax-payer dollars (and citizen stress).

All sorts of management decisions can be similarly related to one another using mathematics.

What is the net gain to a college if it raises tuition by a certain amount? On the one hand, one college will collect more money per student. On the other hand, a number of parents will decide not to send their children to that college. Also, the amount of financial aid given to students will have to be increased to meet the higher tuition rate (which will reduce the net gain to the college). In addition, the cost of recruiting students will increase. To predict the effect of this decision, it would be useful to simulate the situation with a model that reflects the revenues, the expenses, the public-relations impact, the loss of some proportion of applicants, and all the other consequences of the action.

Is it better to manufacture many of one item in one plant distant from some of the markets, or to distribute the manufacturing to various plants? Cost of transportation in this case has to be weighed against the savings made when large amounts are manufactured at one time in one place. A mathematical model that includes all the details of the cost of manufacturing and transportation, including variations by season and by size of order, can answer the question in minutes.

Apart from illustrating the power of mathematics in solving practical problems, these examples also make it clear that, contrary to our grade-school notion that all mathematics produces exact right answers, mathematics seems to be a process of organizing information into categories. Thus, math is quite capable of dealing both with qualitative data and with uncertainty. No wonder familiarity with some of these techniques is useful at the higher reaches of almost every occupation.

Long ago, in Europe, anyone who wanted to participate in government, in centers of learning, and even in certain trades had to learn Latin and sometimes even Greek. Quantitative methods, if not mathematics itself, have become the Latin of the modern era. The vocabulary that derives from systems analysis and the kind of dynamic interactions described in these examples can best be understood by people who have studied math at the level of calculus and beyond. Whether one approves or disapproves of these developments, one has to concede that literacy has a new dimension: mathematical competence.

Notes

1. Frances Rosamond, chair, Department of Mathematics, National University, personal communication.
2. Lynn Osen, "The Feminine Math-tique," Pittsburgh, Pa., K.N.O.W., 1971.
3. U.S. Department of Labor, *Dictionary of Occupational Titles* (Washington, D.C.: U.S. Government Printing Office, 1991).
4. John L. Holland, *Self-Directed Search Inventory* and *The Occupation Finder,* 1989. Available from Psychological Assessment Resources, P.O. Box 998, Odessa, Fla., 33556.
5. Lucy Sells, "Mathematics: A Critical Filter," *Science Teacher,* vol. 45, no. 2 (Feb. 1978).
6. Jean Smith, personal communication.
7. *Peanuts,* Feb. 21, 1979, reprinted with permission of Charles Schulz and United Features.
8. "Expenditures Per Year of Life Saved," *Annual Report,* Harvard School of Public Health, Center for Risk Analysis, May 16, 1991.

2 The Nature of Math Anxiety: Mapping the Terrain

A warm man never knows how a cold man feels.
—Alexander Solzhenitsyn

Symptoms of Math Anxiety

The first thing people remember about failing at math is that it felt like sudden death. Whether it happened while learning word problems in sixth grade, coping with equations in high school, or first confronting calculus and statistics in college, failure was instant and frightening. An idea or a new operation was not just difficult, it was impossible! And instead of asking questions or taking the lesson slowly, assuming that in a month or so they would be able to digest it, people remember the feeling, as certain as it was sudden, that they would *never* go any further in mathematics. If we assume, as we must, that the curriculum was reasonable and that the new idea was merely the next in a series of learnable concepts, that feeling of utter defeat was simply not rational; in fact, the autobiographies of math-anxious college students and adults reveal that, no matter how much the teacher reassured them, they sensed that, from that moment on, as far as math was concerned, they were through.

The sameness of that sudden-death experience is evident in the very metaphors people use to describe it. Whether it occurred in elementary school, high school, or college, victims felt that a curtain had been drawn, one they would never see behind, or that there was an impenetrable wall ahead, or that they were at the edge of a cliff, ready to fall off. The most extreme reaction came from a math graduate student. Beginning her dissertation research, she suddenly felt not only that could she never solve her research problem (not unusual in higher mathematics), but that she had never understood advanced math at all. She, too, felt her failure as sudden death.

Paranoia comes quickly on the heels of the anxiety attack. "Everyone knows," the victim believes, "that I don't understand this. The teacher knows. Friends know. I'd better not make it worse by asking questions. Then everyone will find out how dumb I really am." This paranoid reaction is particularly disabling because fear of exposure keeps us from constructive action. We feel guilty and ashamed, not only because our minds seem to have deserted us, but because we believe that our failure to comprehend this one new idea is proof that we have been "faking math" for years.

In a fine analysis of mathophobia, Mitchell Lazarus explains why we feel like frauds. Math failure, he says, passes through a "latency stage" before becoming obvious either to our teachers or to us. It may actually take some time for us to realize that we have been left behind. Lazarus outlines the plight of the high-school student who has always relied on the memorize-what-to-do approach. "Because his grades have been satisfactory, his problem may not be apparent to anyone, including himself. But when his grades finally drop, as they must, even his teachers are unlikely to realize that his problem is not something new, but has been in the making for years."[1]

It is not hard to figure out why failure to understand mathematics can be hidden for so long. Math is usually taught in discrete bits by teachers who were themselves taught this way; students are then tested as they go along. Some of us never get a chance to integrate all these pieces of information, or even to realize what we are not able to do. We are aware of a lack, but though the problem has been building up for years, the first time we are asked to use our knowledge in a new way, it feels like sudden death. It is not so easy to explain, however, why we take such personal responsibility for having "cheated" our teacher, and why so many of us believe that we are frauds. Would we feel the same way if we were floored by irregular verbs in French?

One thing that may contribute to a student's passivity is the fear of making mistakes in mathematics. Teachers, wanting to reward accuracy, go overboard in treating errors as occasions for *shame,* sure to arouse other students' mirth. Successful math students know better. They do not despise their errors. As one math graduate told a surprised group of math-anxious adults, he finds his mistakes "interesting" because they are "windows into my thinking." Eager to avoid errors at all costs, many children never learn how valuable it would be to explore them. Instead, they just sit in the back of the room hoping the teacher will put those flash cards away.

Another source of passivity is a widespread myth—more common in our culture than in others—that mathematical ability is inborn, and that no amount of hard work can possibly compensate for not having a "mathematical mind." Recent studies of *attitudes* toward mathematical competence in elementary-school pupils, their teachers, and their parents in Japan and Taiwan compared with those in the U.S. show how devastating this myth can be. When asked to explain why some children do better in math than others, Asian children, their teachers, and their parents point to *hard work,* their American counterparts to *ability.*[2]

Leaving aside for the moment the sources of this myth, consider its effects. Since only a few people are supposed to have a mathematical mind, part of our passive reaction to difficulties in learning mathematics is that we suspect we may not be one of "them" and are waiting for our nonmathematical mind to be exposed. It is only a matter of time before our limit will be reached, so there is not much point in our being methodical or in attending to detail. We are grateful when we survive fractions, word problems, or geometry. If that certain moment of failure hasn't struck yet, it is only temporarily postponed.

Sometimes the math teacher contributes to this myth. If the teacher claims to have had an entirely happy history of learning mathematics, she may contribute to the idea that some people—specifically she—are gifted in mathematics, and others—the students—are not. A good teacher, to allay this myth, brings in the scratch paper he used in working out the problem, to share with the class the many false starts he had to make before solving it.

Parents, especially parents of girls, often expect their children to be nonmathematical. If the parents are poor at math, they had their own sudden-death experience; if math was easy for them, they do not know how it feels to be slow. In either case, they will unwittingly foster the idea that a mathematical mind is something one either has or does not have.

Interestingly, the myth is peculiar to math and science. A teacher of history, for example, is not very likely to tell students that they write poor exams or do badly on papers because they do not have a historical mind. Although we might say that some people have a "feel" for history, the notion that one is *either* historical or nonhistorical is patently absurd. Yet, because even the experts don't *really* know how mathematics is learned, we tend to think of math ability as mystical, and to attribute the talent for it to genetic factors. This belief, though undemonstrable, is clearly communicated to us all.

These considerations help explain why math anxiety afflicts such a wide variety of people having diverse mathematical skills. Since we were never "truly mathematical," we had to memorize things we could not understand. And so, with word problems, or with the first bite of algebra, or at the door of calculus (in some cases, even later), the only way we know how to respond to our failure to understand a difficult concept is to quit. Since we never did have a mathematical mind, our act is over, the curtain down.

Ambiguity, Real and Imagined

> What is a satisfactory definition? For the philosopher or the scholar, a definition is satisfactory if it applies to those things and only those things that are being defined; this is what logic demands. But in teaching, this will not do: a definition is satisfactory only if the students understand it.
>
> —H. Poincaré, mathematician and educator

Mathematics autobiographies show that, for the beginning student, the language of mathematics is full of ambiguity. Though mathematics is supposed to have a precise language, more precise than our everyday usage (this is why math uses symbols), mathematical terms are never wholly free of the connotations we bring to words, and these layers of meaning may get in the way. The problem is not that there is anything wrong with math; it is that we are not properly initiated into its vocabulary and rules of grammar.

Some math-disabled adults will remember, after fifteen to thirty years, that the word "multiply" as used for fractions never made sense to them. "Multiply," they remember wistfully, always meant "to increase." That is the way the word was used in the Bible and in other contexts, and surely the way it worked with whole numbers. (Three times six always produced something larger than either three or six.) But with fractions (those with a value of less than one), multiplication always results in something of smaller value. One-third times one-fourth equals one-twelfth, and one-twelfth is considerably smaller than either one-third or one-fourth.

Many words like "multiply" mean one thing (such as "increase rapidly") when first introduced in school. But in the larger context (in this case all rational numbers), the apparently simple meaning becomes confusing. Since students are not warned that "multiplying" has different effects using fractions less than one, they find themselves searching among the meanings of the word to find out what to do. Simple logic, corresponding to the words they know and trust, seems not to apply.

A related difficulty for many math-anxious people is the word "of" as applied to fractions.

In general usage, "of" often means "less than the whole" we're talking about, as in "portion of." Yet with fractions, one-third *of* three-fourths, we were told, is a *multiplication* problem. Rarely are children taught that, in this context, "one-third of" really means "divide by three." If they were, one-third of six and one-third of three-fourths would not seem so different (and might correspond to their real-world experience of slicing pizzas). Barring that correspondence, we have no choice but to suspend our prior associations with the word "of," and to memorize the rule.

Or take the term "least common denominator," part of the rule for adding (and subtracting) fractions. One adult, recalling a particularly painful moment in third grade, took "least common" literally to mean "most uncommon," and was embarrassed to have chosen a very high, instead of the lowest, common denominator to solve her problem. The same ambiguities hold true for symbols. Once we have learned to associate the minus sign with subtraction, it takes an explicit lesson to unlearn the old meaning of "minus" (in connection with negative numbers, for example)—or, as a mathematician would put it, to learn its meaning as applied to another kind of number.

Persis Herold, who offers math tutoring to children and their teachers, has collected examples of this kind of confusion. She reminds us that children are very literal. When a teacher tells them, "You can't subtract 8 from 5," they think you *never* can. But later, with the introduction of the number line (where zero is in the middle, plus and minus numbers on either side), they will be expected to do just this kind of computation (the answer is -3).[3]

Knowles Dougherty, another skilled teacher of mathematics, notes:

> It is no wonder that children have trouble learning arithmetic. If you ask an obedient child in first grade, "What is zero?," the child will call out loudly and with certainty, "Zero is nothing." By third grade, he had better have memorized that "Zero is a

place-holder." And by fifth grade, if he believes that zero is a number that can be added, subtracted, multiplied by, *and* divided by, he is in for trouble.*

People also recall having problems with shapes, never being sure, for example, whether a "circle" is the line around the circle (the perimeter) or the space within. Mathematics is not "fuzzy," mathematicians insist. But it is often taught "fuzzily." Students who experience such difficulties feel they are just dumber than everyone else, but in fact they may be smarter. Mathematicians make just the distinction they seek: the line around is called "the line of points in the plane equidistant from a center." The area of a circle is called the "open disk"† (See figure 2-1).

A mind that is bothered by ambiguity—actual or perceived—is not usually a weak mind, but a strong one. This point is important, because mathematicians argue that it is the learner—not the subject (or its teaching)—who is imprecise. That may be, but since mathematics is often taught to amateurs, differences in meaning between common language and mathematical language do get in our way. It may never be possible to eliminate those differences. As mathematician Sanford Segal says, "No one—not even mathematicians—*thinks* in definition-theorem-proof style." What he and I would argue is that those differences need to be aired so that young people do not blame themselves for their confusions.

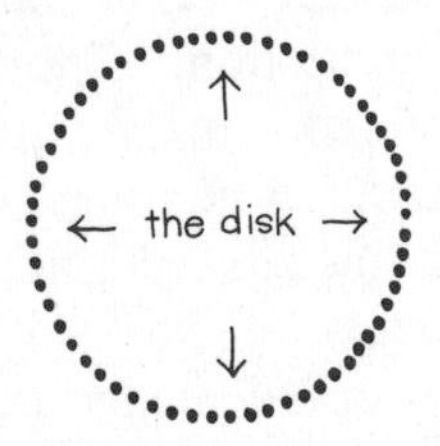

Figure 2-1

The line of points equidistant from the center

*Some of this is changing in more modern elementary texts. But adults coming back to mathematics still have to cope with such leftover confusions.

†My thanks to Sanford Segal for this explanation.

Besides, even if mathematical language is unambiguous, there is no way into it except through our spoken language, in which words are loaded with content and associations. We cannot help thinking "increase" when we hear the word "multiply" because of all the other times we have used that word. We have been coloring circles for years before we get to one we have to measure. No wonder we are unsure of what "circle" means. People who do a little better in mathematics than the rest of us are not as bothered by all this. We shall consider the possible reasons later on.

Meanwhile, the mathematicians withhold information. Mathematicians depend heavily upon customary notation. They have a specific association with almost every letter in the Roman and Greek alphabets, which they don't always tell us about. We think that our teachers are choosing χ or α or delta (Δ) arbitrarily. Not so. Ever since Descartes, the letters at the end of the alphabet have been used to designate unknowns, the letters at the beginning of the alphabet usually to signify constants, and in math, economics, and physics generally Δ, the Greek letter for *d,* means "change" or "difference." Though these symbols appear to us to be chosen randomly, the letters are loaded with meaning for "them."

In more advanced algebra, the student's search for meaning is made even more difficult because it is almost impossible to visualize complex mathematical relationships. For me, the fateful moment struck when I was confronted by an operation I could neither visualize nor translate into meaningful words. The expression $x^{-2} = \dfrac{1}{x^2}$ did me in. I had dutifully learned that exponents such as ^{2} and ^{3} were shorthand notations for multiplication: A number or a letter squared or cubed was simply multiplied by itself once or twice. Trying to translate math into words, I considered the possibility that the expression meant something like "x not multiplied by itself," or "x multiplied by not itself." What words or images could convey to me what x^{-2} really meant? To my question—and I have asked it many times since—the answer is that $x^{-2} = \frac{1}{x^2}$ is a definition consistent with what has gone before. I have been shown examples that demonstrate this consistency.* But at the time, I did not want a demonstration or a proof. I wanted an explanation!

I dwell on the x^{-2} example because I have often asked competent mathematicians to recall for me how they felt the first time they were told $x^{-2} = \frac{1}{x^2}$. Many remember merely believing what they were told in math class, or that they soon found the equivalency useful. Unlike me, they were satisfied with a definition and an illustration that the system works. Why some people should be more distrustful about such matters and less willing to play games of internal consistency than others is a question we shall return to later.

Willing suspension of disbelief is a concept that comes not from mathematics or science but from literature. A reader must give the narrator an opportunity to create images and associations and to "enter" these into our mind (the way we "enter" information into a computer) to carry us along in the story or poem. The very student who can accept the symbolic use of language in poetry where "birds are hushed by the moon," or the disorienting treatment of time in books by Thomas Mann and James Joyce, may balk when mathematics employs familiar words in an unfamiliar way. If willingness to suspend disbelief is specific to some tasks and not to others, perhaps it is related to trust. One counselor explains math phobia by saying, "If you don't feel safe, you won't take risks." People who don't trust math may be too wary of math to take risks.

A person's ability to accept the counterintuitive use of time in Thomas Mann's work and not the new meaning of the negative exponent does not imply that there are two kinds of minds, the

*In multiplying numbers with exponents, we add exponents; in dividing, we subtract. Hence:

$$x^{-2} = x^3 \div x^5 = \frac{x^3}{x^5} = \frac{xxx}{xxxxx} = \frac{1}{xx} = \frac{1}{x^2}$$

The same rule causes $x^0 = 1$.

verbal and the mathematical. I do not subscribe to the simple-minded notion that we are one or the other, and that ability in one area leads inevitably to disability in the other. Rather, I think that verbal people feel comfortable with language early in life, perhaps because they enjoyed success at talking and reading. When mathematics contradicts assumptions acquired in other subjects, such people need special reassurance before they will venture on.

Conflicts between mathematical language and common language may also account for students' distrust of their intuition. If several associated meanings are floating around in someone's head and the text considers only one, the learner will, at the very least, feel alone. Until someone tries to get inside the learner's head or the learner figures out a way to search among the various meanings of the word for the one that is called for, communication will break down, too. This problem is not unique to mathematics, but when people already feel insecure about math, linguistic confusion increases their sense of being out of control. And as long as teachers continue to argue, as they have to me, that words like "multiply" and "of," negative exponents, and "circles" or "disks" are not ambiguous at all but perfectly consistent with their definitions, then students will continue to feel that math is simply not for them.

Some mathematics texts solve the problem of ambiguity by virtually eliminating language. College-level math textbooks are even more laconic than elementary texts. One reason may be the difficulty of expressing mathematical ideas in language that is easily agreed upon. Another is the assumption that by the time students get to college they should be able to read symbols. But for some number of students (we cannot know how many, since they do not take college-level math), proofs, symbolic formulations, and examples are not enough. After I had finally learned that x^{-2} must equal $\frac{1}{x^2}$ because it was consistent with the rule that when dividing numbers with exponents we subtract the exponents, I looked up "negative exponents" in a new high-school algebra text. There I found the following paragraph.

Negative and Zero Exponents

> The set of numbers used as exponents in our discussion so far has been the set of positive integers. This is the only set which can be used when exponents are defined as they were in Chapter One. In this section, however, we would like to expand this set to include all integers (positive, negative and zero) as exponents. This will, of course, require further definitions. These new definitions must be consistent with the system and we will expect all of the laws of exponents as well as all previously known facts to still be true.[4]

Although this paragraph is clear in setting the stage to explain negative exponents through definitions that are presumably forthcoming, it does not provide a lot of explanation. No wonder people who need words to make sense of things give up.

The Dropped Stitch

"The day they introduced fractions, I had the measles," or the teacher was out for a month, or the family moved, or there were more snow days that year than ever before (or since). People who use events like these to account for their failure at math did, nevertheless, learn how to spell. True, math is especially cumulative. A missing link can damage understanding much as a dropped stitch ruins a knitted sleeve. But being sick or in transit or just too far behind to learn the next new idea is not reason enough for doing poorly at math forever after. It is unlikely that one missing link can abort the whole process of learning elementary arithmetic.

In fact, mathematical ideas that are rather difficult to learn at age seven or eight are much easier to comprehend one, two, or five years later if we try again. As we grow older, our facility with language improves; we have many more mathematical concepts in our minds, developed from everyday living; we can ask more and better questions. Why, then, do we let ourselves remain permanently ignorant of fractions or decimals or graphs? Something more is at work than a missed class.

It is of course comforting to have an excuse for doing poorly at math, better than believing that one does not have a "mathematical mind." Still, the dropped-stitch concept is often used by math-anxious people to excuse their failure. It does not explain, however, why in later years they did not take the trouble to unravel the sweater and pick up where they left off.

Fortunately, picking up the dropped stitch won't require you to return to the "Run, Spot, run!" level of arithmetic. A number of new texts, specifically for adults, are on the market.*

When most of us learned math, we learned dependence as well. We needed the teacher to explain, the textbook to drill us, the back of the book to tell us the right answers. Many people say that they never mastered the multiplication table, but I have encountered only one person so far who carries a multiplication table in his wallet. He may have no more skills than the others, but at least he is trying to make himself autonomous. The greatest value of using simple calculators in elementary school may, in the end, be to free pupils from dependence on something or someone beyond their control.

Adults can easily pick up those dropped stitches once they decide to do something about them. In one math counseling session for educators and psychologists, the following arithmetic bugbears were exposed:

> How do you get a percentage out of a fraction like 7/16?
> How do you do a problem like: Two men are painters. Each paints a room in a different time. How long does it take them to paint the room together?
> Where does "pi" come from?†

The issues were taken care of within half an hour.

This leads me to believe that people are anxious not because

*See the Further Reading section for a recommended list.

†See chapter 6 for a discussion of fractions and percents; see chapter 5 for a discussion of the Painting-the-Room Problem. *Pi,* the first letter, *p,* of the Greek word for "perimeter," can be derived by drawing many-sided polygons (like squares, pentagons, hexagons, etc.) and measuring the ratio of their perimeters to their diameters. Even if you do this roughly, the ratios will approach 3.14159 . . .

they dropped a stitch long ago, but, rather, because they accepted an ideology that we must reject: *that if we haven't learned something so far, it is probably because we can't.*

Fear of Being Too Dumb or Too Smart

One of the reasons we did not ask enough questions when we were younger is that many of us were caught in a double bind between a fear of appearing too dumb in class and a fear of being too smart. Why anyone should be afraid of being too smart in math is hard to understand except for the prevailing notion that math whizzes are not normal. Boys who want to be popular can be hurt by this label. But it is even more difficult for girls to be smart in math. Many researchers in the growing field of "gender and education" find that high-achieving college women are especially nervous about competing with men on what they think of as men's turf. This holds for science, engineering, and computer science as well as mathematics. Since many people perceive ability in mathematics as unfeminine, fear of success may well interfere with ability to learn math. And so will fear of taking risks. (See chapter 3.)[5]

The young woman who is frightened of seeming too smart in math must be careful about asking questions in class, because she never knows when a question is a really good one. "My nightmare," one woman remembers, "was that one day in math class I would innocently ask a question and the teacher would say, 'Now, that's a fascinating issue, one that mathematicians spent years trying to figure out.' And if that happened, I would surely

have had to leave town, because my social life would have been ruined." This is an extreme case, probably exaggerated, but the feeling is typical. Mathematical precocity, asking interesting questions, meant risking exposure as someone unlike the rest of the gang.

It is not even so difficult to ask questions that gave the ancients trouble. When we remember that the Greeks had no notation for multidigit numbers, and that even Newton, the inventor of calculus, would have been hard-pressed to solve some of the equations given to beginning calculus students today, we can appreciate that young woman's dilemma.

At the same time, a student who is too inhibited to ask questions may never get the clarification needed to go on. We will never know how many students developed fear of math and loss of self-confidence because they could not ask questions in class. But the math-anxious often refer to this kind of inhibition. In one case, a counselor in a math clinic spent almost a semester persuading a student to ask her math teacher a question *after* class. She was a middling math student, with a B in linear algebra. She asked questions in her other courses, but could not or would not ask them in math. She did not entirely understand her inhibition, but, with the aid of the counselor, she came to believe it had something to do with a fear of appearing too smart.

There is much more to be said about women and mathematics. The subject will be discussed in detail in chapter 3. At this point it is enough to note that some teachers and most pupils of both sexes believe that boys naturally do better in math than girls. Even bright girls believe this. When boys fail a math quiz, their excuse is that they did not work hard enough. Girls who fail are three times more likely to attribute their lack of success to the belief that they "simply cannot do math."[6] Ironically, fear of being too smart may lead to such passivity in math class that eventually these girls develop a conviction that they are dumb. It may also be that these women are not as low in self-esteem as they seem, but by failing at mathematics they resolve a conflict between the neeed to be competent and the need to be liked. The important thing is that, until young women are encouraged to believe that they have the right to be smart in mathematics, no amount of supportive, nurturant teaching is likely to make much difference.

Distrust of Intuition

> Mathematicians use intuition, conjecture and guesswork all the time except when they are in the classroom.
>
> —Joseph Warren, mathematician

> Thou shalt not guess. —sign in a high-school math classroom

Anyone attempting to help adults overcome their math anxiety has to come to grips with *word problems.* Most math-anxiety clinics will feature word problems throughout. In ours, there was always a "weekly word problem" to be solved. Everyone who walked into the clinic, whether a teacher, a math-anxious client, a staff member (including myself), or just a visitor, had to give the word problem a try. Thus, we stimulated numerous experiences with a variety of word problems, and by debriefing both people who solved them with ease and people who gave up on them, we obtained much insight into the way math-anxious people *think* about themselves in relation to math.

One of the arithmetic word problems that we liked to give our students is the Tire Problem.

> A car goes 20,000 miles on a long trip. To save wear, the 5 tires are rotated regularly. How many miles will each tire have gone by the end of the trip?

Most people readily acknowledge that a car has 5 tires and that 4 are in use at any one time. Ordinary math students who are not anxious or blocked will poke around at the problem for a while and then come up with the idea that 4⁄5 of 20,000, which is 16,000 miles, is the answer. They don't always know exactly why they decided to take 4⁄5 of 20,000. They sometimes say it "came" to them as they were thinking about the tires on the car and the tire in the trunk. The important thing is that they *tried* it, and when it resulted in 16,000 miles, they gave 16,000 a "reasonableness test." Since 16,000 seemed reasonable (that is, less than 20,000 miles but not a whole lot less), they were pretty sure they were right.

The math-anxious student responds very differently. The problem is beyond her (or him). She cannot begin to fathom the information. She cannot even imagine how the 5 tires are used (see figure 2-2). She cannot come up with any strategy for solving it. She gives up. Later, in the debriefing session, the counselor may ask whether the fraction 4⁄5 occurred to her at all while she was thinking about the problem. Sometimes the answer will be yes. But if she is asked why she did not try out 4⁄5 of 20,000 (the only other number in the problem), the response will be—and we have heard this often enough to take it very seriously—"I figured that if it was in my head it had to be wrong."

The assumption that if it is in one's head it has to be wrong or,

Figure 2-2

as others put it, "If it's easy for me, it can't be math," is a revealing statement about the self. Math-anxious people seem to have little or no faith in their own intuition. If an idea comes into their heads or a strategy appears to them in a flash, they will assume it is wrong. They do not trust their intuition. Either they remember the "right formula" immediately, or they give up.

Mathematicians, on the other hand, trust their intuition in solving problems and readily admit that without it they would not be able to do much mathematics. The difference in attitude toward intuition, then, seems to be another tangible distinction between the math-anxious and people who do well in math.

The distrust of intuition gives the math counselor a place to begin to ask questions: Why does intuition appear to us to be untrustworthy? When has it failed us in the past? How might we improve our intuitive grasp of mathematical principles? Has anyone ever tried to "educate" our intuition, improve our repertoire of ideas by teaching us strategies for solving problems? Math-anxious people usually reply that intuition was not allowed as a tool in problem solving. Only the rational, computational parts of their brain belonged in math class. If a teacher or parent used intuition at all in solving problems, he rarely admitted it, and when the student on occasion did guess right in class, he was punished for not being able to reconstruct his method. Yet people who trust their intuition do not see it as "irrational" or "emotional" at all. They perceive intuition as flashes of insight into the rational mind. Victims of math anxiety need to understand this, too.

The Confinement of Exact Answers

> Computation involves going from a question to an answer. Mathematics involves going from an answer to a question.
>
> —Peter Hilton, mathematician

Another reason for self-distrust is that mathematics is taught as an exact science. There is pressure to get an exact right answer, and when things do not turn out right, we panic. Yet people who regularly use mathematics in their work say that it is far more useful to be able to answer the question "What is a little more than 5 multiplied by a little less than 3?" than to know *only* that 5 times 3 equals 15. Many math-anxious adults recall with horror the timed tests they were subjected to in elementary, junior, and senior high school with the emphasis on getting a single right answer. They liked social studies and English better because there were so many "right answers," not just one. Others were frustrated at not being able to have discussions in math class. Somewhere they or their teachers got the wrong notion that there is an inherent contradiction between rigor and debate.

This emphasis on right answers has many psychological benefits. It provides a way to do our own evaluation on the spot and to be judged fairly whether or not the teacher likes us. Emphasis on the right answer, however, may result in panic when that answer is not at hand and, even worse, lead to "premature closure" when it is. Consider the student who does get the right answer quickly and directly. If she closes the book and does not continue to reflect on the problem, she will not find other ways of solving it, and she will miss an opportunity to add to her array of problem-solving methods. In any case, getting the right answer does not necessarily imply that one has grasped the full significance of the problem. Thus, the right-answer emphasis may inhibit the learning potential of good students and poor students alike.

In altering the learning atmosphere for the math-anxious, the tutor or counselor needs to talk frankly about the difficulties of doing math. The tutor's scratch paper might be more useful to the students than a perfectly conceived solution. Doing problems afresh in class at the risk of making errors publicly can also link the tutor with the student in the process of discovery. Inviting all students to put their answers, right or wrong, before the class will relieve some of the panic that comes when students fail to get the answer the teacher wants. And, as most teachers know, looking carefully at wrong answers can give them good clues to what is going on in students' heads.

Although an answer that checks can provide immediate positive feedback, which aids in learning, the right answer may come to signify authoritarianism (on the part of the teacher), competitiveness (with other students), and painful evaluation. None of these unpleasant experiences is usually intended, any more than the premature closure or panic, but for some students who are insecure about mathematics, the right-answer emphasis breeds hostility as well as anxiety. Worst of all, the "right answer" isn't always the right one at all. It is only "right" in the context of the amount of mathematics a person has learned so far. We have already seen one example of this: how 8 can't be subtracted from 5 until you learn to deal with negative numbers. But there are many others. First-graders, who are working only with whole numbers, are told they are "right" if they answer that 5 (apples) cannot be divided between 2 (friends). But later, when they work with fractions, they will find out that 5 *can* be equally divided by giving each friend 2½ apples. In fact, both answers are right. You cannot divide 5 one-dollar bills equally between 2 people without getting change.

The search for the right answer soon evolves into the search for the right formula. Some students cannot even put their minds to a complex problem or play with it for a while, because they assume they are expected to know something they have forgotten.

Take this problem, for example:

> Amy Lowell goes out to buy cigars. She has 25 coins in her pocket, $7.15 in all. She has 7 more dimes than nickels, and she has quarters, too. How many dimes, nickels, and quarters does she have?

Most people who have done well in high-school algebra will begin like this: The number of nickels equals x; the number of dimes equals $x + 7$; the number of quarters equals $25 - (2x + 7)$. The total value is $7.15. They won't stop to realize that Amy Lowell must have miscounted her change, because, even if all 25 coins in her pocket were quarters (the largest coin she has), her change would total $6.25, not $7.15.*

This is a tricky problem, which is fair, as opposed to a trick problem, which is not. But it also shows how searching for the right formula can cause us to miss an obviously impossible situation. The right formula may become a substitute for thinking, just as the right answer may replace consideration of other possibilities. Somehow students of math should learn that the power of mathematics lies not only in exactness but in the processing of information as well.

Self-Defeating Self-Talk

One way to show people what is going on in their heads is to have them keep a "math diary," a running commentary of their thoughts, both mathematical and emotional, as they do their homework. Sometimes a tape recorder can be used to get at the same thing. The goal is twofold: to show the student and the instructor the recurring mathematical errors that are getting in the way, and to make the student hear his or her own "self-talk." "Self-talk" is what we say to ourselves when we are in trouble. Do we egg ourselves on with encouragement and suggestions? Or do we engage in self-defeating behaviors that only make things worse?

Inability to handle frustration contributes to math anxiety. When math-anxious people see that a problem is not going to be easy to solve, they tend to quit right away, believing that no amount of time or rereading or reformulation of the problem will make it any clearer. Freezing and quitting may be as much the result of destructive self-talk as of unfamiliarity with the problem. If we think we have no strategy with which to begin work, we may never find one. But if we can talk ourselves into feeling comfortable and secure, we may let in a good idea.

*I am indebted to Jean Smith for this example.

To find out how much we are talking ourselves into failure, we have to begin to listen to ourselves doing math. The tape recorder, the math diary, the self-monitoring that some people can do silently are all techniques for tuning in to ourselves. Most of us who handle frustration poorly in math handle it very well in other subjects. It is useful to watch ourselves doing other things. What do we do there to keep going? How can these strategies be applied to math?

At the very minimum, this kind of tuning in may identify the particular issue giving trouble. It is not very helpful to know that "math makes me feel nervous and uncomfortable" or that "numbers make me feel uneasy and confused," as some people say. But it may be quite useful to realize that one kind of problem is more threatening than another. One excerpt from a math diary is a case in point:

> Here I go again. I am always ready to give up when the equation looks as though it's too complicated to come out right. But the other week, an equation that started out looking like this one did turn out to be right, so I shouldn't be so depressed about it.

This is constructive self-talk. By keeping a diary or talking into a tape recorder, we can begin to recognize our own pattern of resistance, and with luck we may soon learn to control it. This particular person is beginning to understand how and why she jumps to negative conclusions about her work. She is learning to sort out the factual mistakes she makes from the logical and even the psychological errors. Soon she will be able to recognize the mistakes she makes *only* because she is anxious. Note that she has been encouraged to think and to talk about her feelings while doing mathematics. She is not ashamed or guilty about the most irrational of thoughts, not frightened to observe even the onset of depression in herself; she seems confident that her mind will not desert her.

The diary or tape-recorder technique has only been tried so far with college-age students and adults. As far as we can tell, it is effective only when used in combination with other nonthreatening teaching devices, such as acceptance of discussion of feelings in class, psychological support outside of class, and an instructor willing to demystify mathematics. The goal in such a situation is not to get the right answer. The goal is to achieve mastery and, above all, autonomy in doing math. In the end, we can only learn when we feel in control.

Notes

This chapter is based primarily on interviews with and observations of math-anxious college students and adults, including their recollections of learning math in the lower grades. I have also benefited from discussion with instructors and math-anxiety clinicians: Frances Rosamond of National University; Kathryn Brooks and Ashley DuLac of the University of Utah; Susan Auslander and Bonnie Donady of the Wesleyan Math Clinic; Jean Smith, now teaching adults in Maine; Cindy Arem of Pima Community College; and Gary Horne, Mary Ellen Hunt, and Pamela Reavis of Tucson. Knowles Dougherty, Peter Hilton, Robert Rosenbaum, Mitchell Lazarus, and Stanley Kogelman guided my earlier thinking on the subject. Sanford Segal of the University of Rochester was my major adviser for the new edition.

1. Mitchell Lazarus, "Mathophobia: Some Personal Speculations," *Principal,* Jan.–Feb. 1974, p. 18.
2. Harold W. Stevenson and James W. Stigler, *The Learning Gap* (New York: Summit Books, 1992). See also National Center for Educatio Statistics, *The Nation's Report Card:* (Washington, D.C.: Department of Education, 1992).
3. Persis Herold of the Math Learning Center, Washington, D.C., personal communication.
4. Louis Nanney and John L. Cable, *Elementary Algebra: A Skills Approach* (Boston: Allyn and Bacon, 1974), p. 215.
5. Matina Horner, "Fear of Success," *Psychology Today,* Nov. 1969, pp. 38 ff. For an update, see Gilah Leader, "Math Achievement and Fear of Success," *Journal for Research in Math Education,* vol. 13 (1982), pp. 124–35.
6. This is dealt with in chapter 3, under "attribution theory."

3 Mathematics and Sex: Perceptions and Reality

In January 1989, a national report on mathematics education included for the first time the following statement about mathematics and sex: "Gender differences in mathematics performance are predominantly due to the accumulated effect of sex-role stereotypes in family, school, and society."[1] That may be the *official view* informed by twenty years of research and advocacy, but it is still not the *popular view*. In the fifteen years since *Overcoming Math Anxiety* was first published, many stereotypes that once held sway have crumbled. Textbooks no longer mock adult women's attempts at simple calculation. Larry no longer paints the room faster than Susie or earns more than she does for helping Mr. Todd (see figure 3-1). Television often features a woman scientist or mathematician. Teachers and other professionals are careful to say "she" as well as "he." But the myth of a "male math gene" persists, and career data show women still avoiding math and the math-based fields in disproportion to men.

Of the 4.6 million U.S. citizens employed in science and engineering, 30 percent are women, but the vast majority are in the life sciences and psychology, the less mathematical of these careers. Only 4 percent are employed in engineering, the same percent in physics—the most mathematical of the sciences. Men, as I wrote in 1978, are not free to avoid mathematics; a majority of women still believe they are.

In reviewing the vast number of studies of women and mathematics, which I did for this edition, I believe that the reality of sex differences in mathematical ability is still unknown. Some males—those at the very high end of mathematical performance—continue to outdistance outstanding females (though not by much).[2] But *ability* can never be measured absolutely. No one has yet identified a "math lobe" in the brain. Researchers can only measure performance on tests, and we know much more now than we used to about how much performance is influenced by beliefs, perceptions, prior experience, and self-esteem. The experts differ, the debate goes on. But since mathematics re-

If Larry earns $3.94 a day for helping Mr. Todd, how much does he earn in 3 days? in 4 days?

Susie also helps Mr. Todd after school. She earns $1.49 a day. How much money does Susie earn in 4 days? in 7 days?

Figure 3-1

mains in the popular mind the bottom line of sex differences in intelligence, the question that needs to be answered is why.

We know that there are differences in *interest* in mathematics between the sexes. We are only beginning to know what causes such differences, and, even more important, whether they are innate or learned. Do girls do poorly in math because they are afraid that people (especially boys) will think them abnormal if they do well, or is it because girls are not taught to believe that they will ever *need* mathematics? Do girls do certain kinds of math better than boys? Which kinds? At what ages? And are there different ways to explain key concepts of math that would help some girls understand them better?

In 1974, John Ernest, a professor of mathematics at the University of California at Santa Barbara, put mathematics and sex on the public's agenda for the first time. Assigned to teach a freshman seminar about elementary statistics, Ernest decided to turn his seminar into an investigation of the relationships, real and imagined, between gender and performance in mathematics. His students fanned out into neighboring junior and senior high schools to interview teachers and students about girls' and boys' performance in mathematics. The results of their inquiry were nearly always the same. Both boys and girls, they were told, have a fair amount of trouble doing math, and most of them do not like the subject very much. The difference between them was that boys stuck with math, because they felt their careers depended on it and because they had more confidence than girls in their ability to learn it. The problem, concluded Ernest, was not math inability in females; it was their math *avoidance* at crucial stages of their schooling. Society expects males to be better than females at mathematics. This affects attitudes; attitudes affect performance; performance affects willingness to study more mathematics; and, eventually, males do better than females. His findings, modest though they were, were considered so important that the Ford Foundation published and distributed forty thousand copies of his little book.[3]

In fact, math avoidance is not just a female phenomenon. Most people of both sexes stop taking math well before their formal education is complete. Few Americans become mathematicians, and many smart people do not like math at all.* Thus, "dropping out" of math is nearly universal and is by no means restricted to girls. From this perspective, girls who avoid math and math-related subjects may simply be getting the message sooner than boys that math is unrewarding and irrelevant, but boys will also get that message in time.

We see this in *The Nation's Report Card,* a regularly appearing survey of math/science (and other) achievement of America's young people. Few performance differences between males and females at ages nine or thirteen appear, but consistent though small differences in favor of males appear at age seventeen.[4] The bulk of those differences show up in the top 10 to 20 percent of the students who take these tests;[5] but—and this is rarely reported in the popular press—the differences between boys and girls (between-sex differences) are never as large as the differences among girls and among boys (within-group differences).[6] (See figure 3-2.) What ought to matter more to Americans seeking to regain our economic and technological stature is that the greatest differences in mathematics performance are not between American boys and American girls, but between American young people and nearly all the other young people in the world![7]

Still, sex differences in performance and, more important, perceptions that mathematics is a "male domain," persist. Fewer U.S. females than males enroll in advanced mathematics courses in high school such as trigonometry, precalculus, and calculus. The most mathematical of the sciences—physics—is least populated by female students. Fathers, children report, are more likely to help with math homework than mothers.[8] Even teachers, expecting more in mathematics of their boy students than of their

*In 1960, only 3.9 percent of college graduates majored in math. By 1985, this had declined to 1.7 percent. Of these, between 40 and 50 percent will be females. That's because girls who are interested in math generally major in math; boys with the same aptitude choose engineering, computer science, the physical sciences, accounting, and management. Thus, female math majors constitute pretty much *all* the college women interested in math, male math majors only a small percentage of the total.

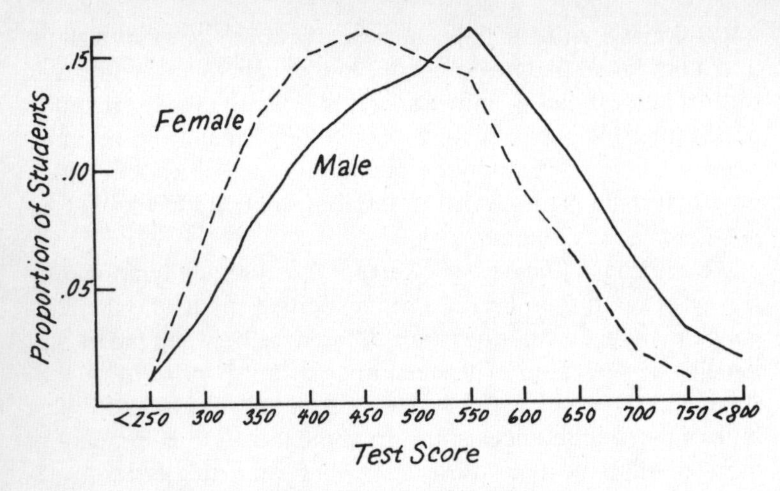

Figure 3-2a SAT Math

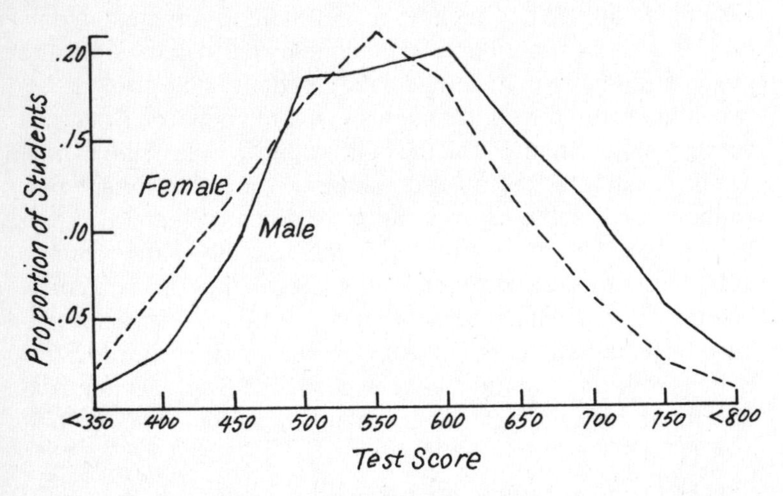

Figure 3-2b

It's important to realize that the mathematics performance of average males and females mostly overlaps. Figure 3-2a denotes scores on the SAT math-aptitude test; figure 3-2b scores on the SAT mathematics I achievement test in a typical year. *Courtesy of The College Board.*

girls, ask higher-order questions of boys and encourage males to discover alternative solutions while inhibiting girls' mathematical creativity by insisting that they follow the rules.[9]

There is no question that boys are more exposed to mathematical thinking at home, in preschool, and in the games they play. Preschool boys more than girls are given blocks and construction toys—whether they ask for them or not—which develop spatial skills (see chapter 4). Older boys receive more home training in handling tools, machines, electrical repairs, and cars—skills that enhance their spatial sense and further ready them for laboratory subjects.[10] Now that the computer has made its way into middle-class homes, experience on the computer gives boys another edge. A 1985 Stanford University study showed that middle- and upper-class boys, grades five to eight, were three times more likely to have access to home computers than girls. Not surprisingly, three-quarters of the high-school-student population taking elective computer courses continues to be male.[11]

At the end of schooling, writes Elizabeth Fennema in her introduction to a survey of mathematics and gender published in 1990, there should be no differences in what females and males have learned, nor should there be any differences in how students feel about themselves as learners of mathematics. "Males and females should be equally willing to pursue mathematics-related careers and should be equally able to learn new mathematics, as it is required."[12] This is, however, not yet the case. Is it because, as some researchers would have us believe, there are sex differences in mathematics ability that no amount of equal treatment can eliminate? Or are there *internal* (though not innate) as well as *external* variables that cause girls not to do as well as boys? My own research, along with that of Fennema and many others, compels me to focus on issues we overlooked in the first wave of a feminist critique of mathematics and sex: (1) female isolation in the math world; (2) the findings of attribution theory; (3) a new insight—Fennema's contribution to the discussion in the 1990s—that females do not develop autonomous learning behavior in mathematics—that is, the all-important willingness to take risks; and (4) the continuing notion that one is good either in mathematics or in the language arts, causing even mathematically able girls to dismiss and disparage their talent.

Female Isolation

It took me a long time to recognize that the crucial question in assessing girls' attitudes and experience in mathematics was not the simple "Do you like math?" or "Do you like your math teacher?" or "Do you like your math class?" but, rather, "With whom and how often do you *discuss* what you're learning in math?" To the first questions, we often got very positive responses. Indeed, as I described in chapter 2, many people remember liking math in school until they hit a wall. But to the crucial question about *math conversation,* the answers were generally "With no one" and "Never." Unless they are blessed with a math-oriented family, or collected in a special dormitory for math/science majors,* girls find themselves isolated both in class and outside of class when it comes to math. Not having anyone to talk to about what they're learning, they fail to learn to *speak mathematics;* worse yet, they do not get the opportunity to extend their knowledge, their skills, and their imagination through discussion.

Much has been written about girls' need for adult role models, for women who are succeeding in or, better yet, really enjoying their work in math-based fields. But far less attention has been paid to age-peers. One exception is the work of Uri Treisman, a mathematician and mathematics educator who has won deserved kudos for identifying isolation as a factor in the failure of able African-Americans to do well in college calculus at the University of California at Berkeley. In a much-quoted study, using the methods of anthropology, Treisman had field workers compare what Asian-American and African-American students from the same calculus course did in their after-class time. The Asian-

*More and more available at colleges like Douglass (at Rutgers University, in New Jersey) and Pennsylvania State University.

Americans organized themselves into "study gangs," working together not only on their homework, but also challenging themselves with harder-than-textbook problems. The African-Americans at Berkeley had been so isolated in their high schools that they had not developed group study habits. And this, Treisman concluded, was the reason they were not doing as well.

To test his hypothesis, Treisman created "academic excellence" groups that, instead of providing remedial help for the weaker students, gave them, by design and under supervision, the essence of what the Asian-Americans were able to give themselves: the stimulus of an out-of-class environment in which mathematics was talked about and practiced. The technique worked, and since 1987 has been initiated in dozens of campuses around the country.[13]

The movie *Stand and Deliver,* released in 1988, showed how a class of mathematically disadvantaged inner-city students could master advanced-placement calculus when goaded by an able and challenging teacher. Jaime Escalante, a computer scientist turned high-school teacher, did not want to accept the prevailing view that calculus is a course only for "geniuses." He contracted with his class of thirty seniors (half of them females) to discipline themselves and to do considerably more than average homework for a year. In exchange, he coached them to the success he made them believe was within their capacity.

Much of the attention *Stand and Deliver* received, and of subsequent discussion of the film, has focused on Escalante the teacher. But at least as much credit must be laid at the door of Escalante the team-builder. For what he intuited was the value of group identity and collaboration in tackling a difficult subject.

Mathematics, more than other subjects, is taught to students in isolation from one another. Flash cards, timed tests, and frequent exposure at the blackboard intensify that isolation and encourage competition. One of our math-anxious undergraduates, in retelling his math autobiography, characterized math classes as being too private and too public at the same time. He had to do his thinking privately at his desk; he was not allowed to speak or ask questions or discuss anything with his neighbor or think out loud. And then he had to expose his work, embarrassed though he might be by it, to the entire class.

Girls, the new research reveals, are just as isolated as Treisman's African-Americans, even earlier in their math careers. Cheating paranoia, on the part of teachers, reduces the amount of peer interaction for both sexes in class. Sex stereotyping makes even the best-performing girls unwilling to reveal how much they like mathematics; hence, they talk less about mathematics out of class. The result may be reduced learning by girls, who are highly verbal to begin with. Isolation, the experts tell us, leads to alienation, a sense of not belonging in a particular subject area. Minority females suffer this isolation even more.

Attribution Theory

Since 1978, most researchers have shifted their focus from "innate differences" to the influence of *choice behavior* on what women students do.[14] Girls' choices, as already noted, are linked to what they perceive to be the *value* of mathematics in their lives. But they are also linked to girls' *expectations* of how well they will do if they were to continue in math. Why should young girls, generally more successful at school in the earliest years, not expect to continue to do as well? Why do even the ablest girls not derive heightened confidence from their early success in mathematics? Attribution theory gives us some clues.

Boys and girls, we are told, differ significantly in the way they explain to themselves (and to the researchers who interview them) both their current and past successes and failures in mathematics.[15] Most boys, as table 3-1 illustrates, attribute their success in math to ability; girls attribute theirs to effort. Conversely, most boys explain their failure on a particular math task or test by the fact that they didn't work hard enough. Girls take the low grade (even if it's the only low one they get in an entire year) to mean they're just not smart enough for math.

Why might this be? Gary Horne, a mathematics teacher currently teaching math to young adults in the juvenile justice sys-

	SUCCESS	FAILURE
Boys	ability	(not enough) effort
Girls	effort (or sometimes luck)	(not enough) ability

Table 3-1

How explanation of success and failure at mathematics varies among high school students by sex

tem, observes that disempowered groups in general—and not just women and girls—attribute failure to lack of ability.[16] Effort, he speculates, may appear to be irrelevant in a world in which certain groups already feel alienated. Success, of course, is the trickier issue. How does a student attribute success if few who resemble him or her are successful in a particular subject? Is it a fluke? Is it effort? As a result of disenfranchisement, concludes Horne, success simply may not even be recognized as success.

The girl who attributes her math success to consistent effort rather than ability, says psychologist Jacquelynne Eccles, will have low expectations for future success precisely because she thinks her future courses will be even more difficult and demanding than the ones in which she is currently enrolled. Women's lack of confidence about math is compounded by failure and not enhanced by success. This may be why girls do not enroll at the same rate as boys in more advanced math classes in high school and college, and why they become more negative and pessimistic about their math skills as they grow older. Clearly, confidence in and enjoyment of mathematics go hand in hand. Fifty percent of children at age nine say that math is their favorite subject; by age seventeen, the number who like math best is down to less than 20 percent, and girls "precede the boys in this downward trend."[17]

For a while, after the math-anxiety issue was first raised, researchers failed to find a direct causal link between math anxiety and career choice among women. But in recent years, as the nation's interest in a possible "shortfall" of science workers has emerged, researchers have gone back to see whether attitudes toward mathematics and toward likely success at mathematics correlate with women students' willingness to consider science as a career.[18] In a 1991 survey of Barnard (an all-women's college) students who scored well above average on precollege quantitative tests—that is, whose "math ability" ought to have inclined them toward science—25 percent indicated that their previous achievements in mathematics did not translate either into a desire to continue or into increased confidence. As these highly able students revealed to the researchers, it was their "desire to avoid mathematics" that was primarily affecting their choice of careers.[19]

How else might we explain a resistance on the part of girls and women to *enjoy* and *take credit* for their achievements in math?

Autonomous Learning Behavior

In 1985, mathematics educator Elizabeth Fennema and her colleagues proposed the "autonomous learning behavior" model as a possible explanation for the persistence of gender differences in mathematics. A student who is autonomous, said Fennema, increasingly assumes *control* of her learning. She *chooses* to engage in high-level mathematical tasks and prefers to work *independently* on them.[20] When the task proves difficult, the autonomous learner *persists*. Such behaviors, say the experts, will certainly lead to success in mathematics and, as a result of this success, to habituation (what psychologists call "reinforcement") of the behaviors themselves.

What is attractive about Fennema's idea is that, although it is not gender-specific, it goes far to explain why girls have less confidence than boys about higher math. Both sexes have the capacity for autonomy and independence. We know that. But we also know that little (and not-so-little) girls are often punished when they display too much moxie. There is nothing innately male either about succeeding in mathematics or about having a high tolerance for risks; our society, in one more effort to keep women down, simply declares these to be masculine. The difference between *male* (a biological category) and *masculine* (a societally produced category) is the difference between *nature* and *nurture* in that age-old debate. But the question remains: How do we teach our girls and our women to *be independent,* to *take risks,* and to want to *be in control?*

"You only take risks when you feel safe," Bonnie Donady would begin her math-anxiety sessions for women. And well she might. In study after study, girls are shown to be far more concerned with what other people think of them, unhappy with problems that have more (or fewer) than one right answer, unwilling to guess or speculate, and apparently needing more feedback than boys as to how they are doing.[21] If readiness to do word problems, to take one example, is as much a function of readiness to take risks as it is of "reasoning ability," then there is more to mathematics performance than memory, computation, and thought. The differences between boys and girls—no matter how consistently they show up—cannot simply be attributed to differences in innate ability.

Good Either in Math or in Language Skills

There is a myth—as exaggerated and misleading as most myths are—that one is either "good in English" or "good in math and science," but rarely in both. In fact, mathematicians write very well and usually score high on the verbal sections of standardized tests. What is different about thinking mathematically is that it may sometimes be more efficient to work with symbols than with words, and to write equations instead of sentences. The *encoding* and *decoding* of mathematical symbols is not substantially different from writing and reading. Yet youngsters, and sometimes their parents, too, think that these involve very different talents and skills.

There is lots of evidence that little girls speak sooner, read earlier, and have generally more language fluency in the earlier grades than boys. Fewer girls than boys stutter, have dyslexia (perceptual difficulties), or suffer other blocks in reading. Thus, little girls have a head start and learn to *trust* words even more than they trust the information that comes in through their eyes and their hands.[22]

Mathematics, as we have already noted, is a kind of language, and words are as much a part of math lessons as they are of geography. Arithmetic facts have to be packaged in language; teacher talks and students listen. And when children get to word problems (see chapter 5), understanding what is being asked for—vocabulary and *interpretation* of words and sentences—is the first step toward solving the problem.

Still, when we ask victims themselves, people who have trouble doing math, they do not believe their language skills will help them master math. They believe their disability has to do with the way their brains are "wired." They feel they somehow lack something—one ability or several—that other people have. Although women want to believe they are not mentally inferior to men, many fear that in math they really are. Thus, we must consider seriously how much "socialization" can influence behavior in the face of persistent arguments from biology.

How Can You Recognize "Sexism" in Mathematics?

The psychologists and mathematics educators who have propounded some of the theories referred to in this chapter have employed complex (and carefully validated) tests of attitudes and behaviors in conjunction with mathematics performance. But you can do some simple self-monitoring exercises to see whether prejudice or sex-role expectations are blocking you (or anyone you know) from achieving mathematical competence. Imagine that you get a higher-than-usual grade on a particular math test. Which of the following reflects your explanation for this feat?

1. The content included things I already knew.
2. I spent hours of extra time studying math.
3. The way the teacher presented the material helped.
4. I am a better math student than I thought I was.[23]

If you choose any of these attributions but (4), you may be having trouble taking *ownership* of your success. Why might this be so? Sex stereotyping is everywhere we look.

A talking Barbie doll recently distributed (and just as quickly removed because of protest) was programmed to say, "Math class is tough," when a child turned her on.[24] Why do you think the manufacturers didn't have her say, instead, "I plan to be a mathematician when I grow up"? Would there have been more laughs than sales of Barbies? Or would she have been not quite so cute? What do you think is the *latent* (subliminal) message in figure 3-3 (adapted from an elementary textbook that we hope is no longer in use)? Or in figure 3-4? Why have you (or most of the men you know) never heard of Hypatia, one of the greatest mathematicians (male or female) of antiquity; Sonya Kovalevskaya, who took nineteenth-century mathematics by storm; Emmy Noether, considered by far the most creative algebraist who ever lived; Grace Murray Hopper, who invented the computer languages most businesses use today; or Lise Meitner, a mathematical physicist who figured out that the chemists in her lab had split the uranium atom. It was Meitner who named what happened "fission."[25]

Why aren't women of mathematics as much celebrated as men? Partly because, until recently, they have been few in number, and partly because their gender is considered either grounds for exclusion or not pertinent at all. The title of a many-times reissued classic history of mathematics is *Men of Mathematics,* first published in 1937, two years after the "most creative algebraist" (Professor Noether) had died. One might forgive the original author for his oversight, but why reissue the book again in 1965 without a title change?[26] The deepest reason, I think, lies in the way the culture views mathematics itself. Because it is stripped of detail, removed from nature, many practitioners think of mathematics not as a tool but as something akin to *pure thought.* How can ordinary women, rooted in the everyday, ascend to such heights? The fact is, in most cultures (not just our own), women have been unwelcome as serious intellectuals bearing skills that men want to believe are theirs alone. Once women are excluded, men are able to convince themselves that women are less talented than they are. The tragedy is that *intellectual* men, men of science, have participated in that charade. More tragic still is that men of science continue to invoke arguments from biology even today.

The Arguments from Biology

The search for some biological basis for math ability or disability is fraught with logical and experimental difficulties. Since not all math underachievers are women and not all women avoid mathematics, it is not very likely on the face of it that poor performance in math can result from some genetic or hormonal difference between the sexes. Moreover, no amount of research has so far unearthed a "mathematical competency" in some tangible, measurable substance in the body. Since masculinity cannot be injected into women to see whether it improves their mathematics, the theories that attribute such ability to genes or hormones must depend on circumstantial evidence for their proof. To explain the percent of Ph.D.'s in mathematics earned by women, we would have to conclude either that these women have different genes, hormones, and brain organization than the rest of us, or that certain positive experiences in their lives have largely undone the negative influence of being female, or both.

Londa Schiebinger, a historian of women in science, tells us that our nineteenth-century forebears were so eager to find women biologically inferior to men that they first asserted that the lighter weight of women's brains accounted for their lesser cognitive abilities and power. Later, when it became obvious that the *ratio* of women's brain weight to their total weight was higher than that of men, the pseudo-scientists changed their argument. The ratio of children's brain to body weight is even higher; hence, women, like children, were "dumber" than men.[27]

Many other claims were made in the past, most of them thoroughly discounted today. (See Stephen Jay Gould's *Mismeasure*

Figure 3-3

Figure 3-4

of Man.[28]) Just about the time women were demanding higher education, doctors pronounced that their reproductive systems competed for limited energy with their mental activities. So, during their childbearing years (most of their adult lives), they ought not to exercise their minds—that is, attend college—lest this result in "puny wombs."

At the root of many of the assumptions about biology and intelligence is the undeniable fact that there have been fewer women "geniuses." The distribution of genius, however, is more a social than a biological phenomenon. An interesting aspect of the lives of geniuses is precisely their dependence on familial, social, and institutional supports. Without schools to accept them, men of wealth to commission their work, colleagues to talk to, and wives to do their domestic chores, they might have gone unrecognized—they might not even have been so smart. In a classic essay explaining why we have so few great women artists, Linda Nochlin Pommer tells us that women were not allowed to attend classes in art schools because of the presence of nude (female) models. Nor were they given apprenticeships or mentors; and even when they could put together the materials they needed to paint or sculpt, they were not allowed to exhibit their work in galleries or museums.

Women in mathematics fared little better. Emmy Noether, who, we have already noted, was considered a genius, was honored (or perhaps mocked) during her lifetime by being called "Der" Noether ("Der" being the masculine form of "the"). Der Noether notwithstanding, the search for the genetic and hormonal origins of math ability goes on.

Genetically, the only difference between males and females (albeit a significant and pervasive one) is the presence of two chromosomes designated "X" in every female cell. Normal males have an "X-Y" combination. Since some kinds of mental retardation are associated with sex-chromosomal anomalies, a number of researchers have sought a link between specific abilities and the presence or absence of the "X." But the link between genetics and mathematics is simply not supported by conclusive evidence.

Since intensified hormonal activity begins at adolescence and since, as we have noted, girls seem to lose interest in mathematics during adolescence, much more has been made of the unequal amounts of the sex-linked hormones, testosterone and estrogen, in males and females. Estrogen is linked with "simple repetitive tasks" and testosterone with "complex restructuring tasks." The argument here is not only that such specific talents are biologically based (probably undemonstrable), but also that such talents are either-or: that one cannot be good at *both* repetitive and restructuring kinds of assignments.

Further, if the sex hormones were in any way responsible for our intellectual functioning, we should get dumber as we get older, since our production of both kinds of sex hormones decreases with age.* But as far as we know, hormone production responds to mood, activity level, and a number of other external and environmental conditions as well as to age. Thus, even if one day we were to find a sure correlation between the amount of hormone present and the degree of mathematical competence, we would not know whether it was the mathematical competence that caused the hormone level to increase or the hormone level that gave us the mathematical competence.

All this criticism of the biological arguments does not imply that what women do with their bodies has no effect on their mathematical skills. As we will see, toys, games, sports, training in certain cognitive areas, exercise, and experience may be the intervening variables we have previously mistaken for biological cause.

*Indeed, some people do claim that little original work is done by mathematicians once they reach age thirty. But a counterexplanation is that creative work is done not because of youth but because of "newness to the field." Mathematicians who originate ideas at twenty-five, twenty, and even eighteen are benefiting not so much from hormonal vigor as from freshness of viewpoint and willingness to ask new questions. I am indebted to Stewart Gillmor, historian of science, for this idea.

Sex Hormones and Brain Organization

Today's arguments from biology are, superficially, more sophisticated than they once were. In a special issue of *Scientific American* published in September 1992 and featuring "The Mind and the Brain," researcher Doreen Kimura argues that sex hormones not only influence our sex drive, our reproductive function, and our moods, but also determine the way we solve intellectual problems. Worse yet, as she writes, "the effects of sex hormones on brain organization occur so early in life that from the start the environment is acting on differently wired brains in girls and boys."[29] Her findings are particularly destructive of any claim that women and men are equal (and alike) because she finds differences in *patterns of intellectual ability* and not just in general intelligence—precisely the kind of argument that buttresses the claim that men are better in one mental arena, women in another.

Men, on average, she argues, perform better than women on certain spatial tasks—an issue we shall return to in chapter 4. They outperform women in mathematical-reasoning tests and in navigating their way through a route. They are more accurate in target-directed motor skills—that is, in guiding or intercepting projectiles. Women, on the other hand, have greater perceptual speed—that is, they more rapidly identify matching terms, have greater verbal fluency, and outperform men in arithmetic calculation and in recalling landmarks from a route.[30] Worst of all (from my point of view), these differences appear before puberty, the result of different sex-hormone levels present at or shortly after birth.

How do Kimura and her colleagues know this? Since they cannot manipulate the hormonal environment in humans (we have "human-subjects" protection, which makes that illegal), they have to deprive male and female rats of certain hormones artificially and then observe their behavior. Clearly, the "problem-solving skills" they can measure in rats are considerably different from the problem-solving skills used by humans to do math. So, when they observe female rats (or testosterone-deprived male rats) using "landmark cues" such as pictures on a wall to orient themselves, and male rats using geometric cues like the angles and the shape of the room, the researchers draw some pretty large conclusions: first, that these are significant variations; second, that they are the result of structural differences in the brain; and third, that they can be extended to explain differences in math ability between human males and human females.[31]

How can they be so sure? All they can do is observe "parallel" behavior in humans, such as girls' and boys' styles of play, and the more tomboyish behavior of girls whose mothers were given androgens during pregnancy. I cannot see how a rat's finding her way out of a maze can possibly correlate with higher-order problem-solving skills in human beings. In studies like these, it usually turns out that there is a smaller or larger brain segment in one sex or the other. Presto! A physical explanation for male superiority in mathematics. Kimura's measurements are more complex, her arguments more subtle, but we have not come as far as we thought from nineteenth-century phrenology, in which (it was thought) the number of bumps on the head predicted intellectual prowess.

This is not to say that neuroscience has nothing to tell us about the human brain, or about how humans evolved with, possibly, sex-differentiated brain organization. The division of labor required of our ancestors in primitive hunter-gatherer societies may well have favored long-distance route-finding among males, and different but equally important skills among females. But to leap from all this speculation to the claim that today's men and women may have different occupational interests and capabilities (engineering for men, Kimura tells us, medical diagnostics for women) as a *result* of their biological sex is, I think, pure bunkum. Yet it is leaps like these from small measurable differences to gross social implications that catch us off our guard. The responsible neuroscientist has no business speculating, especially when writing for a general public, because the public only remembers the speculations and is in no position to argue the physiological detail.

The Benbow-Stanley Furor

Just such a wild speculation characterized the publication in 1980 of a paper co-authored by Julian Stanley and Camilla Benbow, then researchers at Johns Hopkins University. Because of that paper, our campaign to persuade women (and men) that, if properly encouraged, girls could do just as well in mathematics as boys was set back perhaps a dozen years. The paper appeared, along with much hoopla directed toward the popular press, in *Science,* a journal of repute and wide readership among scientists. Based on tests of ten thousand self-selected mathematically precocious eleven-year-olds, the report conveyed the authors' conviction that "large sex differences" in mathematical *aptitude* were found in boys and girls who had essentially the same formal educational experiences and got approximately the same grades in math in school.[32]

Again, as in many other studies, the largest sex differences appeared among the top-scoring boys and girls—sometimes as high as 190 points out of a possible 800; and, on average, boys outnumbered girls in the group that scored above 500. Though the authors admitted that a *liking* for mathematics might predict mathematics *achievement* in the long run, they concluded in no uncertain terms that mathematical *aptitude* is the result of "superior male mathematical ability."[33]

It didn't take long for the notion that there is a "male math gene" to make its way into the popular press (see figure 3-5). "A new study says males may be naturally abler than females" in mathematics, reported *Time* magazine to a readership that would not have the patience to wade through Benbow and Stanley's tables of data or to read the criticism that the editors of *Science,* in an effort to counter the effect of the article, summarized in an editorial in the same issue. If they had, they would have realized that fewer girls enroll in the accelerated math courses because, as one interviewer found, they were afraid of being labeled "different" by their friends; also because the talented boys were "little creeps."[34] But the damage was done. Another generation of girls and their parents would accept lower performance in math than in other subjects.

Much has happened in the years since Benbow and Stanley first burst upon the "mathematics-and-sex" scene with their controversial results. Benbow has gone on to correlate other biological factors (left-handedness, tendency to allergies, and myopia) with mathematical precocity. She has also carefully followed up the "graduates" of the Johns Hopkins program through their career lives. Stanley continues to study gender differences in standardized tests, finding male superiority in most achievement areas except spelling, clerical speed, and accuracy.[35] He is surprised that feminists "deplore" his work. Perhaps it's because he rejects out of hand the notion that girls' ability to think under pressure (that is, during tests) might be influenced by the way their parents, their teachers, and their friends define the "female role."

Sex Roles and Mathematics Competence

A group of seventh-graders in a private school in New England gives us a clue as to how early female sex roles are learned. When they were asked why girls do as well as boys in math until sixth grade, but after sixth grade boys do better, the girls responded: "Oh, that's easy. After sixth grade, we have to do real math." These young girls have already bought into what Lynn Osen calls the "feminine mathtique." Parents, peers, and teachers will forgive them if they ever do badly in math at school, encouraging them to do well in other subjects instead.

" 'There, there,' my mother used to say when I failed at math," one woman remembers. "But I got a talking-to when I did badly in French." "Mother couldn't figure out a 15-percent tip, and Daddy seemed to love her more for her incompetence," remembers another. Lynn Fox, who worked with Benbow and Stanley on the Johns Hopkins program for mathematically gifted children, found it difficult to recruit and keep girls in the program.

Do Males Have a Math Gene?

Can girls do math as well as boys? All sorts of recent tests have shown that they cannot. Most educators and feminists tude Test normally given to high-school seniors. In the results on the math portion of the SAT—there was no appreciable dif-

Newsweek, Dec. 15, 1980

The Gender Factor in Math

A new study says males may be naturally abler than females

Until about the seventh grade, boys and girls do equally well at math. In early high school, when the emphasis Julian C. Stanley of Johns Hopkins University, males inherently have more mathematical ability than females.

Time, Dec. 15, 1980

Male superiority

Are boys born superior to girls in mathematical ability? The answer is probably Yes, say Camilla Persson Benbow and Julian C. Stanley, researchers in the department of psychology at the Johns

The Chronicle of Higher Education, December, 1980

Are Boys Better At Math?

New York Times, Dec. 7, 1980

BOYS HAVE SUPERIOR MATH ABILITY, STUDY SAYS

Boys are inherently better at math than girls, according to an eight-year study of 10,000 gifted students. Coun-

Education U.S.A., Dec. 15, 1980

SEX + MATH = ?

Why do boys traditionally do better than girls in math? Many say it's because boys are encouraged to pursue

Family Weekly, Jan. 25, 1981

Study suggests boys may be better at math

WASHINGTON (UPI) – Two psychologists said Friday boys are better than girls in math reasoning, and they urged educators to accept the fact that something more than social factors is re-

Ann Arbor News, Dec. 6, 1980

Figure 3-5 Headlines following the Benbow-Stanley Studies

Taken from Jacquelynne S. Eccles and Janis E. Jacobs, "Social Forces Shape Math Attitudes and Performance," *Signs: Journal of Women in Culture and Society,* vol. 11, no. 2 (Winter 1986), pp. 367–80. See also Doris K. Yee and Jacquelynne S. Eccles, "Parent Perceptions and Attributions for Children's Math Achievement," *Journal of International Education,* 1988, pp. 317 ff.

Their parents sometimes prevented them from participating altogether; they claimed not to have "noticed" their daughters' mathematical gifts. The girls themselves often paid for their participation with social ostracism. The math-anxious college student we met in chapter 2 who would have lost her social standing if she had asked an interesting question in math class was anticipating just that.

The fact is, extreme conformance to socially defined sex roles may not serve either gender well. Research shows that intelligence in general (and possibly even mathematical intelligence would fit this model) correlates not with extreme masculinity or femininity but with cross-sex identification. Boys and girls who pursue some of the interests and behaviors of the opposite sex score higher on general-intelligence tests and tests of creativity than children who are exclusively masculine or feminine.[36] Girls who resist the pressure to become ladylike and instead develop aggressiveness, independence, self-sufficiency, and tough-mindedness score higher on these tests than more passive, "feminine" girls. And, similarly, boys who are "sensitive" score higher than more typical, aggressive boys.

Average boys may feel the need to misbehave in school because they are getting mixed messages from their parents: be naughty on the playing field but quiet and controlled in school. Yet girls are expected to be consistently docile. Another possibility is that children of both sexes who are more intelligent to begin with may find society's rules cumbersome and choose to ignore them. That is, either intelligence breeds behavior that is not typical of the child's sex, or unconventional experience stimulates intelligence, or both.

There is far less information about personality characteristics related to mathematical performance than about personality and intelligence in general. But one recent study compared three groups of women college students, one group higher in verbal abilities than in math, the second group higher in math than in verbal, and the third group about equally competent in both. The study found that the women higher in math ability responded positively to a cluster of attributes considered masculine, such as "logical," "persistent," and "intellectual." But this group also scored high on positively valued feminine attributes, such as "warm," "generous," and so on. The researcher concluded that these women were not rejecting femininity itself, but only such low-valued feminine characteristics as dependence and passivity. The women seemed to have a healthy orientation toward the best of both the male and female worlds.

Such a conclusion is an important modification of past studies that found women who do well in math to be "masculine." One such study, linking problem-solving ability in math with a masculine self-image, even went so far as to conclude that nonmathematical men had an image problem. Another early analysis of the autobiographies of women in mathematics concluded that these women either lacked a "typical feminine identification" or were "conflicted" over their female role. Far better and more recent studies of women mathematicians do not explain their success by their masculine or feminine "nature," but find that these women enjoy some real, tangible advantages, among them strong family support.

If sex-role socialization is what one is taught about oneself by others, then we may call what one learns about oneself by oneself "experience." And we must look into the impact of *experience* on learning math.

One appealing theory sounds almost too simple. It is that people who do well in mathematics from the beginning and people who have trouble with it have altogether different experiences in learning math. These differences are not necessarily innate or cognitive or even, at the outset, differences in attitude or in appreciation for math. Rather, they are differences in how people cope with uncertainty, whether they can tolerate a certain amount of floundering, whether they are willing to take risks, what happens to their concentration when an approach fails, and how they feel about failure. These attitudes could be the result of the kinds of risks and failures they remember from early experience with mathematics, because our expectations of ourselves are shaped not simply by what others say but by what we think we can and cannot do.

Street Mathematics: Things, Motion, and Scores

If a ballplayer is batting .203 going into a game (fifteen hits in seventy-four times at bat) and gets 3 hits in 4 times at bat (which means he has batted .750 for the day), someone watching the game might assume that the day's performance will make a terrific improvement in his batting average. But it turns out that the 3-for-4 day only raises the season average of .203 to .231. Disappointing, but a good personal lesson in fractions, ratios, and percents.

Scores, performances like this one, lengths, and speeds of sprints or downhill slaloms are expressed in numbers, in ratios, and in other comparisons. The attention given to such matters surely contributes to a boy's familiarity with simple arithmetic functions and must convince him, at least on some subliminal level, of the utility of mathematics. This does not imply that every boy who handles runs batted in and batting averages well during the game on Sunday will see the application of these procedures to his Monday-morning math assignment. But handling figures as people do in sports probably lays the groundwork for using figures later on.

Not all the skills necessary for mathematics are learned in school. Measuring, computing, and manipulating objects that have dimensions and dynamic properties of their own are part of everyday life for some children. Children who miss these experiences may not be well primed for math in school.

Feminists have complained for a long time that playing with dolls is one way to convince impressionable little girls that they may only be mothers or housewives, or, in emulation of the Barbie doll, pinup girls when they grow up. But doll playing may have even more serious consequences. Have you ever watched a little girl play with a doll? Most of the time she is talking and not doing, and even when she is doing (dressing, undressing, packing the doll away) she is not learning very much about the world. Imagine her taking a Barbie doll apart to study its talking mechanism. That's not the sort of thing she is encouraged to do. Do girls find out about gravity and distance and shapes and sizes playing with dolls? Probably not!

A college text written for inadequately prepared science students begins with a series of supposedly simple problems dealing with marbles, cylinders, poles made of different substances, levels, balances, and an inclined plane. Even the least talented male science student will probably be able to see these items as objects, each having a particular shape, size, and style of movement. He has balanced himself or some other object on a teeter-totter; he has watched marbles spin and even fly. He has probably tried to fit one pole of a certain diameter inside another, or used a stick to pull up another stick, learning leverage. Those trucks little boys clamor for and get are moving objects. Things in little boys' lives drop and spin and collide and even sometimes explode.

The more curious boy will have taken apart a number of household and play objects by the time he is ten; if his parents are lucky, he may even have put them back together. In all this, he is learning things that will be useful in physics and math. Taking out parts that have to go back in requires some examination of form. Building something that stays up or at least stays put for some time involves working with structure. Perhaps the absence of things that move in little girls' childhoods (especially if they are urban little girls) quite as much as the presence of dolls makes the quantities and relationships of math alien to them.

In sports played, as well as sports watched, boys learn more math-related concepts than girls do. Getting to first base on a not very well hit grounder is a lesson in time, speed, and distance. Intercepting a football in the air requires some rapid intuitive eye calculations based on the ball's direction, speed, and trajectory. Since physics is partly concerned with velocities, trajectories, and collisions of objects, much of the math taught to prepare a student for physics deals with relationships and formulas that can be used to express motion and acceleration.* A young woman

*In learning physics, unlike math, however, students must unlearn intuitive notions: for example, heavier objects will not always fall faster. So, to some extent, real-life experience may be counterproductive.

who has not closely observed objects travel and collide cannot appreciate the power of mathematics.*

Unfamiliarity with things may also cause a girl to distrust her environment. Since the movement of objects seems not only irregular but capricious, watching things move may not seem to her to be as reliable a way to learn about the world as following the lesson in a book. A wilderness canoe instructor confirms this as he describes a woman learning to canoe:

> When I start my preliminary instruction, she hangs on to my words, watching me intently. When she gets into the canoe, she mimics exactly what I have done, even if it is inappropriate. . . . She wants to know how to put the paddle in the water, how hard to pull on it, when to start pulling, where to hold it. . . . She makes the operation into a ritual like a dance, becoming increasingly tense (and frustrated) as time goes on. . . .
>
> At some point I say to her, "Now look, you are trying to get your feedback from the wrong place. You keep watching me and you should be watching the boat. The boat will tell you what is right and what is wrong. When you do it right the boat does what you want it to.
>
> "I can tell you what to look for, but I can't tell you how it feels. As long as you keep watching me you will get nowhere. Now forget about being instructed. Just go out in the canoe and play around with it and find out what it does."[37]

The point is that what we get out of an experience, even a good one, may depend on what we have done and learned before. One thing is what you know you *can* do; another is what you think you *should* do; and the combination of limited physical experience and negative attitudes toward math may be the principal contributor to females' poorer performance in mathematics.

*Even the arithmetic games that girls like to play will hardly teach them about the world of natural physical events. Monopoly and playing store provide practice in arithmetic fundamentals, but nothing that might suggest some of the more complex phenomena in her environment that mathematics can explain.

Conclusion

After surveying the summaries of research in this area and interviewing people who claim to be incompetent at mathematics, I have reached a conclusion. Apart from general intelligence, which is probably equally distributed among males and females, the most important elements in determining success at learning math are motivation, temperament, attitude, and interest. These are at least as salient as genes and hormones (about which we really know very little in relation to math), "innate reasoning ability" (about which there is much difference of opinion), or number sense. This does not, however, mean that there are no sex differences at all.

What is ironic (and unexpected) is that, as far as I can judge, sex differences seem to be lodged in *acquired skills:* not in computation, visualization, and reasoning *per se,* but in ability to take a math problem apart, in willingness to tolerate certain kinds of ambiguity, and in careful attention to mathematical detail. Such temperamental characteristics as persistence and willingness to take risks may be as important in doing math as pure memory or logic. And attitude and self-image—particularly during adolescence, when the pressures to conform are at their greatest—may be even more important than temperament. Negative attitudes, as we all know from personal experience, can powerfully inhibit intellect and curiosity and can keep us from learning what is well within our power to understand.

An Afterword

Does "math anxiety," particularly when applied to women, imply that women are more impressionable and weaker in spirit than men? Several feminists criticize the anxiety model, pointing out that, since the causes of math anxiety lie in "political and social forces that oppress women" and are not wholly psychological and educational in origin, the goal of remediation should not be "the curing of an individual case but the elimination of the conditions that foster the disease."[38]

The identification of mathematics anxiety as a problem for women could become two-edged. Focusing on one more female "disability" may feed the prejudices that already abound in the real world about women and math, women and science, and women and machines. We also have to consider the needs of women who are very competent in math and have a hard time proving this to their colleagues. Finally, we have to contemplate the possibility that attention given to this issue might expose women to exploitation by "math-anxiety experts."

Mindful of all these objections, I still argue that excessive anxiety inhibits women more than it does men. If you ask any female how she feels about mathematics, you may find out how she feels about many other gender-related aspects of her life as well. Very bright girls who excel at almost everything in school feel quite comfortable failing at math, not simply because their parents allow it and their peers accept it, but because it provides a solution to the conflicts their brightness creates for them. Rebellious adolescent girls, on the other hand, may actually force themselves to like and do well at math as a way of holding their femininity at bay for a while.

Trying to help one young woman graduate student overcome her intense hostility to math, Stanley Kogelman, cofounder of Mind over Math, heard her say that it was the "logic" and "discipline" of mathematics that she disliked most. Probing to find out where those feelings about "logic" and "discipline" came from, Kogelman concluded that the woman was really disturbed by the fear that she would enjoy the rigorous part of her *own* mind. Mathematics was incidental in her struggle. She was actually in conflict over her own identity.[39]

Do men suffer from math anxiety, and does it intrude as much in their lives as it does in women's? Until we have more satisfactory measures of math anxiety and have more math autobiographies from men, we will not have much to say about this. But some researchers believe that, although men have math anxiety, too, it doesn't trouble them quite as much. One study was done with 655 Ohio State University undergraduates enrolled in a pre-calculus course. The researcher tested their math anxiety (as best she could with a paper-and-pencil questionnaire) and then compared their anxiety ratings with their final grades. She found, interestingly, that the men's math-anxiety scores did *not* correlate with their final course grades nearly as much as the women's anxiety scores correlated with theirs.[40] Perhaps the men went to greater lengths to hide their anxiety even from the researcher. Or perhaps, as the researcher concludes, math anxiety is harder for women to overcome.

My hunch is that the researcher is right. Men have math anxiety, too, but it disables women more.[41]

Notes

There are many books for the concerned parent determined to prevent or alleviate math anxiety and to counter myths about "girls and mathematics" that children (of both sexes) may bring home from school. Among my favorites are Teri Perl, *Math Equals: Biographies of Women Mathematics and Related Activities* (Menlo Park, Calif.: Addison-Wesley, 1978); Jean Kerr Stenmark, Virginia Thompson, and Ruth Cossey, *Family Math* (in English) *Mathematica Para La Familia* (in Spanish) (Berkeley, Calif.: Lawrence Hall of Science, 1986), a treasure trove of things for the whole family to do together in math; and Joan Skolnick, Carol Langbort, and Lucille Day, *How to Encourage Girls in Mathematics: Strategies for Parents and Educators* (Englewood Cliffs, N.J.: Prentice Hall, 1982).

1. *Everybody Counts: A Report to the Nation on the Future of Mathematics Education* (Washington, D.C.: National Academy Press, 1989).
2. Gilah Leder, "Gender Differences in Mathematics: An Overview," in Elizabeth Fennema and Gilah Leder, eds., *Mathematics and Gender* (New York: Teachers College Press, 1990), p. 13.

3. John Ernest, *Mathematics and Sex* (New York: Ford Foundation, 1976). Also published in *American Mathematics Monthly,* vol. 83, no. 8 (Oct. 1976), pp. 599 ff.
4. Data is from the 1983 National Assessment of Educational Progress quoted in Gilah Leder, "Gender Differences in Mathematics: An Overview," in Fennema and Leder, eds., *Mathematics and Gender,* p. 13. High-school seniors taking the SAT quantitative-aptitude test show males with a forty- to fifty-point advantage over girls, largely because girls take fewer quantitative science courses in high school. See C. Navarro, "Why Do Women Have Lower Average SAT-Math Scores Than Men?" paper presented at the American Educational Research Association Convention, 1989, ERIC Document Reproduction Service No. ED 312-277.
5. These findings are true in Great Britain as well. See L. Joffe and D. Foxman, "Attitudes and Sex Differences: Some APU Findings," in L. Burton ed., *Girls into Maths Can Go* (London: Holt, Rinehart and Winston, 1986).
6. Gilah Leder, "Gender Differences in Mathematics: An Overview," in Fennema and Leder, eds., *Mathematics and Gender,* p. 13.
7. Harold W. Stevenson and James W. Stigler, *The Learning Gap* (New York: Summit Books, 1992), Chap. 3.
8. Jacquelynne Eccles (Parsons) et al, "Self-Perceptions, Task-Perceptions, Socializing Influences, and the Decision to Enroll in Mathematics," in S.F. Chipman, L.R. Brush, and D.M. Wilson, eds., *Women and Mathematics: Balancing the Equation* (Hillsdale, N.J.: Lawrence Erlbaum, 1985).
9. E. Fennema and P.L. Peterson, "Effective Teaching, Student Engagement in Classroom Activities, and Sex-related Differences in Learning Mathematics," *American Educational Research Journal,* vol. 22, no. 3 (1985), pp. 309–35.
10. Janet Lever, "Sex Differences in Games Children Play," *Social Problems,* vol. 23, no. 4 (April 1976), pp. 79–90.
11. Lynn Alper and Meg Homberg, *Parents, Kids, and Computers* (Berkeley, Calif.: Lawrence Hall of Science, 1984).
12. Elizabeth Fennema, "Justice, Equity, and Mathematics Education," in Fennema and Leder, eds. *Mathematics and Gender,* p. 5.
13. Uri Treisman, described the "Emerging Scholars Program" at the University of Texas during a Chautauqua Short Course via Satellite, April 8, 1992.

14. Eccles (Parsons) et al., "Self-Perceptions," pp. 95 ff.
15. This section draws on a large body of work in attribution theory, beginning with B. Weiner, ed., *Achievement Motivation and Attribution Theory* (Morristown, N.J.: General Learning Press, 1974), as summarized and enlarged upon by Eccles (Parsons) et al., "Self-Perceptions," pp. 101–03.
16. Gary Horne, personal communication.
17. Eccles (Parsons) et al., "Self-Perceptions," pp. 101–03.
18. Susan F. Chipman, David H. Krantz, and Rae Silver, "Mathematics Anxiety and Science Careers Among Able College Women," *Psychological Science,* vol. 3, no. 5 (Sept. 1992), pp. 292 ff.
19. Ibid.
20. Margaret Meyer and Mary Schatz Koehler, "Internal Influences on Gender Differences in Mathematics," in Fennema and Leder, eds., *Mathematics and Gender,* p. 69.
21. Eleanor Maccoby and Carol Nagy Jacklin, *The Psychology of Sex Differences* (Stanford, Calif.: Stanford University Press, 1974).
22. Again, it is only performance and preference for talking that can be measured, not "verbal ability." Although observation indicates that girls talk earlier and are more fluent than boys, even feminist researchers dispute that this represents a higher "ability" in girls. See Anne Fausto-Sterling, *Myths of Gender: Biological Theories About Women and Men* (New York: Basic Books, 1985), pp. 26–30.
23. Adapted from E. Fennema et al., "Mathematics Attribution Scale." For a discussion, see Peter Kloosterman, "Attributions, Performance Following Failure, and Motivation in Mathematics," in Fennema and Leder, eds., *Mathematics and Gender,* pp. 122–25.
25. Kevin Sullivan, "Foot in Mouth Barbie: Talking Doll's Patter Irks Math Teachers," *Washington Post,* Sept. 30, 1992. This Barbie was pulled off the market one month later.
25. For biographies of women mathematics and exercises for high-school students introducing them to the mathematics in which these women pioneered, see Teri Perl, *Math Equals,* op. cit. A scientific biography of Lise Meitner by Ruth Sime will be published by the University of California Press in 1995.
26. E. T. Bell, *Men of Mathematics: The Lives and Achievements of The Great Mathematicians from Zeno to Poincare* (orig. ed. 1937; New York: Simon and Schuster, 1965).
27. Londa Schiebinger, *The Mind Has No Sex: Women in the Origins of Modern Science* (Cambridge, Mass.: Harvard University Press, 1989).

28. Stephen Jay Gould, *Mismeasure of Man* (New York: W.W. Norton, 1980).
29. Doreen Kimura, "Sex Differences in the Brain," *Scientific American,* Sept. 1992, pp. 121 ff.
30. Ibid., p. 120.
31. Ibid., p. 121.
32. Camilla P. Benbow and Julian C. Stanley, "Sex Differences in Mathematical Ability: Fact or Artifact?," *Science,* vol. 210 (Dec. 12, 1980), pp. 1262–64. Their study of mathematically precocious youth took place from 1972 through 1979.
33. Ibid., p. 1264.
34. The interviewer was Lynn Fox, then associated with the program, as quoted in Gina Bari Kolata, "Math and Sex: Are Girls Born with Less Ability?," *Science,* vol. 210 (Dec. 12, 1980), p. 1235.
35. Julian C. Stanley et al., "Gender Differences on Eighty-six Nationally Standardized Aptitude and Achievement Tests," paper presented at the Wallace Research Symposium, University of Iowa, May 17, 1991.
36. Eleanor C. Maccoby and Carol Nagy Jacklin, *Development of Sex Differences* (Stanford, Calif.: Stanford University Press, 1966), p. 43.
37. John D. McRuer, canoeist and instructor for the Algonquin Waterways Wilderness Tours, Toronto, Canada. Letter to author, 1976.
38. Patti Hague, "Technology Assessment from a Feminist Perspective: Math Anxiety Programs," unpublished article, 1976. See also Judith Jacobs, "Women and Mathematics: Must They be at Odds?," *Pi Lambda Theta Newsletter,* Sept. 1977.
39. Stanley Kogelman, "Debilitating Mathematics Anxiety: Its Dynamic and Etiology," unpublished master's thesis, Smith College School of Social Work, 1975.
40. Nancy E. Betz, "Math Anxiety: What Is It?" paper presented at the American Psychological Association Convention, San Francisco, 1977.
41. For further information on the subject of this chapter, see the author's "Math Anxiety: Why Is a Smart Girl Like You Counting on Your Fingers?," *Ms.,* Sept. 1976, pp. 56 ff; *Succeed with Math: Every Student's Guide to Conquering Math Anxiety* (New York: College Board, 1987); with Bonnie Donady, "Counseling the Math Anxious," *Journal of the American Women Deans, Counselors, and Administrators,* Fall 1977, p. 13; with Bonnie Donady, "Math Anxiety," *Teacher,* Nov. 1977, p. 71.

    The Women's Educational Equity Act Publishing Center at 55

Chapel St., Newton, Mass., 02130, distributes articles, videos, and "sex-fair" materials on girls and mathematics and science. For a free catalog call toll free 1-800-225-3088, or in Massachusetts, 617-969-7100. The video, "Math Anxiety: We Beat It So Can You" a film about the Wesleyan Math Anxiety Program, is available from that source.

4 Right- and Wrong-headedness: Is There a Nonmathematical Mind?

Spatial Visualization

As we have seen, not all women experience math anxiety, and not all people who fear mathematics are women. This is essential to the proper understanding of the preceding chapter. There is, however, one knotty facet of intelligence that prevents researchers from abandoning altogether the study of male/female differences in intellectual activities. Tests of the ability to understand and manipulate drawings of two-dimensional and three-dimensional figures generally show males to be more skilled in the area than females. This skill is called "spatial visualization." Even among very intelligent girls and women, the capacity to visualize shapes moving through space is less well developed than in comparably intelligent males. When large groups are compared, males on the high end do better than comparable females at finding embedded figures hidden in complex drawings, solving geometric problems, "cutting" cubes, learning mazes, and reading maps. Researchers have even found that telling left from right varies to some extent by sex. Try out your own spatial visualization on the tests beginning with figure 4-1.

The first example in two dimensions requires a simple rotation. The goal is to find the image in the collection A, B, C, D, E that is exactly the same as the one in the example. In each example the image has been rotated, but the letters must be in the same relation for a correct answer. If you have to "translate" the operation into terms like these:

> The stem of the *T* faces the right-hand vertical of the *H*, and the *H* is in the very next corner to the *T*; therefore, the only items among the group of possible answers where these two conditions are met are B, C, and D. Since there are three possibilities, I need to look also at the *F*. Okay, the long side of the *F* faces the brace of the *H*. This eliminates C among the answers, and also D. The answer must be B. (And it is.)

then you are tackling the problem verbally and not spatially. To do it spatially, you should be able just to etch the image of the one into your mind and find the right answer by mentally rotating the image. Try figure 4-2.

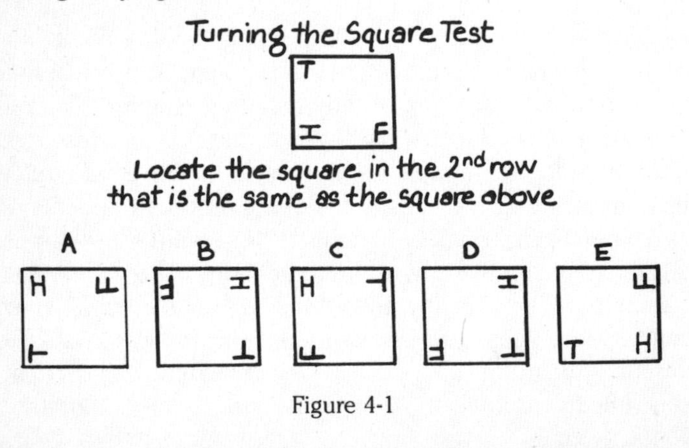

Figure 4-1

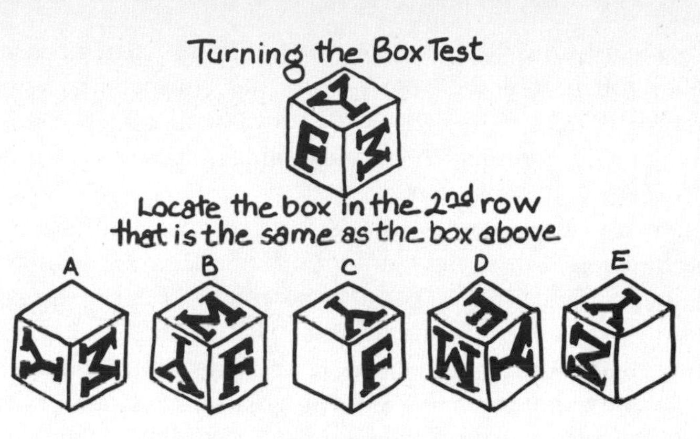

Figure 4-2

Here the item is in three dimensions, but the rotation operation is similar. Again, if you have to talk out the relationship between the stem of the *Y* and the side of the *M* and note that the three invisible sides of the cube face the long stem of the *F*, the right side of the *Y*, and the top side of the *M*, you are verbalizing the test. The correct answer is not A because the stem of the *Y* meets the bottom side of the *M*. B is incorrect because, when the *M* is on top, the *F* is upside down. C is not correct because the *Y* and the *F* are in the wrong relation. D is right.

Try another item involving flipping as well as rotation (figure 4-3). The correct answer for this test item is d. But if you had to say to yourself, as I did, "The example is moved clockwise one-quarter of a turn," you are coping verbally, not spatially, with the problem.

I think it is obvious that, in any more complicated items than

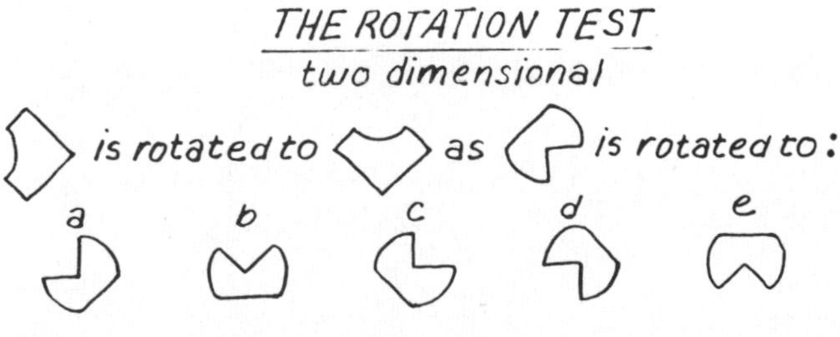

Figure 4-3

these, the rotations, flips, and turns would be virtually impossible to talk through. Hence, at some point in such a spatial-skills test, either we would see the right answer or we would have to stop.

Try not talking through the next examples (figures 4-4, 4-5, and 4-6). (See p. 130 for answers.)

There are very few measurable sex differences in intelligence. Therefore, much has been made of this one factor, spatial intuition, and particularly of its possible bearing on analytic thinking. Since there is even speculation that poor spatial visualization alone can account for difficulty in learning advanced mathematics, anyone trying to understand the greater incidence of math avoidance and math anxiety in women must consider the possibility that their poorer spatial skills are the critical factor. But is there sufficient evidence that all or most women do less well than men? And can we definitely link the ability to see spatial relations and math? These are the critical issues.

During World War II, women were not entirely trusted as trained pilots. Although some women did fly freight missions (and Amelia Earhart had previously won trophies for her even more daring flights), the reason given for not training more women as pilots was that they were not as good as men at spatial visualization. They tended not to be able to right themselves while blindfolded and strapped to a tilted chair in a tilted room.*

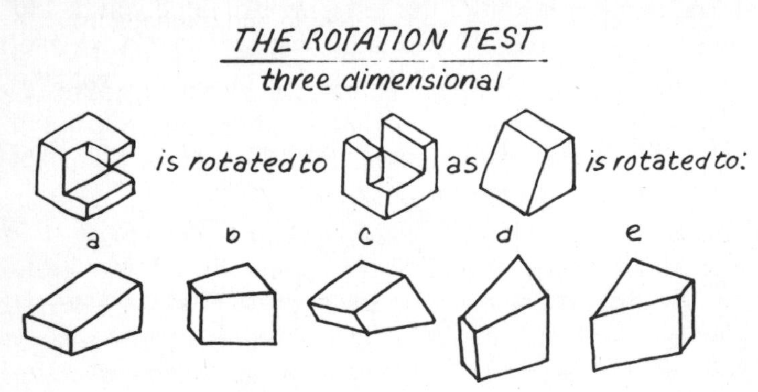

Figure 4-4

*The so-called rod-and-frame test.

Finish the Box Test

Find the piece or pieces that complete the figure on the left

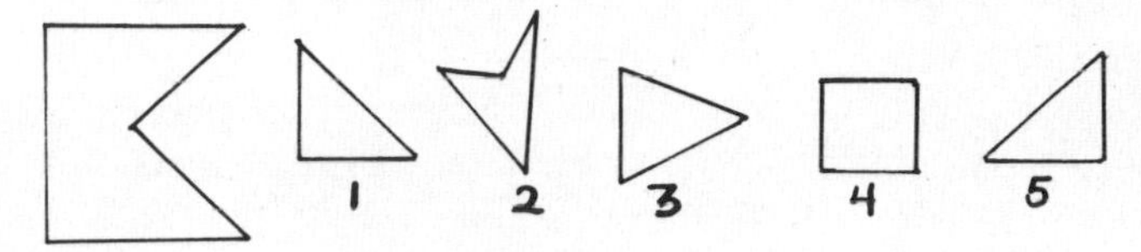

Figure 4-5

Fold the Box Test

Find the folded box that represents the unfolded box on the left

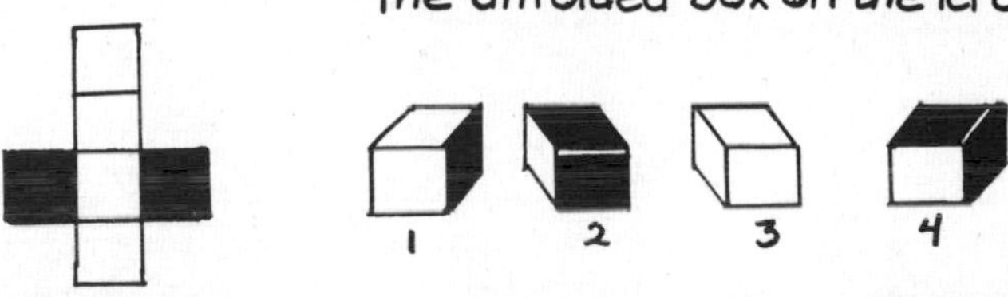

Figure 4-6

Many other occupations have also used some form of spatial-skills testing on an entrance examination and, perhaps unintentionally, as a way to exclude women. Even now, most dental schools test spatial-visualization ability not by having applicants carve a plaster cast, as was once required, but with a paper-and-pencil test including items similar to the ones we have just examined.

This is one misuse of the spatial-ability test, since there is no firm evidence that persons who do poorly on the test do not become competent dentists. Another misuse is in the assumed connection between analytic ability—a much more widely appreciated talent than mere spatial ability—and spatial visualization. A frequently used measure of analytic ability is the "Embedded-Figure Test."

In this test (figure 4-7), the goal is to find, among an assortment

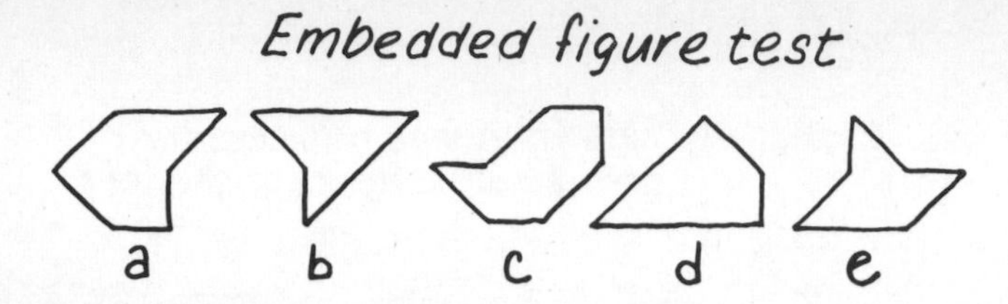

INSTRUCTIONS:
Four of the five shapes above are concealed in one or more of the boxes below. Which shapes are in which boxes?

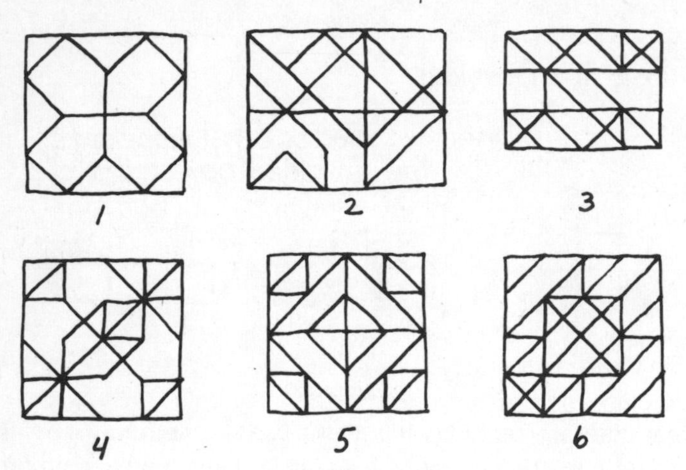

Figure 4-7

of items, the one in which the shape given at the outset is embedded. Since the shape is mixed with an array of confusing lines, what is supposedly being tested is the person's ability to sort out the item from its context, the essential characteristic from its confusing environment. The problem with this test as a measure of analytic ability, as Julia Sherman has pointed out, is that the characteristic is not an idea or a theme or an inconsistency—the sort of things a good analyst must sort out from confusing data—but, rather, a *shape.* Sherman argues powerfully that the Embedded-Figure Test should not be used, because it measures two "abilities," spatial visualization and analytic capability, not one. Thus, it is unfair to women, who are not as good as men at spatial visualization. On other tests of analytic ability, where the item to be located is a key sentence in a paragraph or a number out of place in a series, women do as well as men.

But if sex differences in analytic ability do not exist (as most of us hope and believe), there is still some evidence that spatial skills seem less well developed in women (on the average) than in men. Where does spatial intuition come from? Is it lodged in the brain, as some researchers think? Or is it learned, much as boys learn to manipulate objects and scores through their play, their physical environment, and their expectations of themselves?[1]

The Theory of the Two Hemispheres

Whether spatial visualization can be learned and improved, and whether this skill is related to mathematics, are issues that have been debated for a long time. But one area of research, on the structure and organization of the brain itself, is relatively new. Researchers now think that they can locate spatial visualization in the brain and can explain why certain women and men do not have well-developed spatial faculties.[2]

During the past few decades, several independent lines of research have been generating a new model of the brain. In this model, the brain consists of two relatively independent information processors—a right hemisphere and a left hemisphere—that function in different ways. Research is usually done on brain-damaged people: stroke victims who are paralyzed on one side of their bodies, and epileptics whose brain hemispheres have been split surgically to prevent the seizures from spreading from one part of the brain to the other. By giving these patients specific intellectual tasks to perform, neuroscientists have established the duality of the right and left hemispheres. The right hemisphere of the brain, which dominates the left side of the body, seems to contribute the abilities to perceive shapes, to remember musical phrases, to grasp things (though not necessarily ideas) as wholes, and to recognize faces. The left hemisphere, which dominates the right side of the body, specializes in speech and other linear or sequential tasks. Letters have to follow a certain order to make words, words must follow a certain order to make sentences, and the ability to grasp these sequences seems to be lodged in the left hemisphere. A person whose left hemisphere has been damaged by a stroke cannot speak or move the right side of his body. Yet he may well be able to sing and remember lyrics. Conversely, if the right hemisphere is malfunctioning, the patient will be able to read and write but may not recognize faces or remember geographical locations.

Nevertheless, people who have had a hemisphere disconnected, through either brain damage or surgery, can do both right-hemisphere and left-hemisphere tasks. But they may not be able to use with one side of the brain information that went into the other. One startling example is the patient who, with her right eye covered, gives an embarrassed giggle when a nude figure flashes before her left eye but cannot explain why she laughs.[3]

The particular disabilities of the severed brain were not anticipated when this kind of surgery was introduced. Neurosurgeons expected to find a wall of tissue between the two hemispheres of the brain. Instead, they found a billion "telephone lines," each linking some special function in one half of the brain to a function in the other. From this physical evidence, doctors came to believe that cross-brain communication is much more important

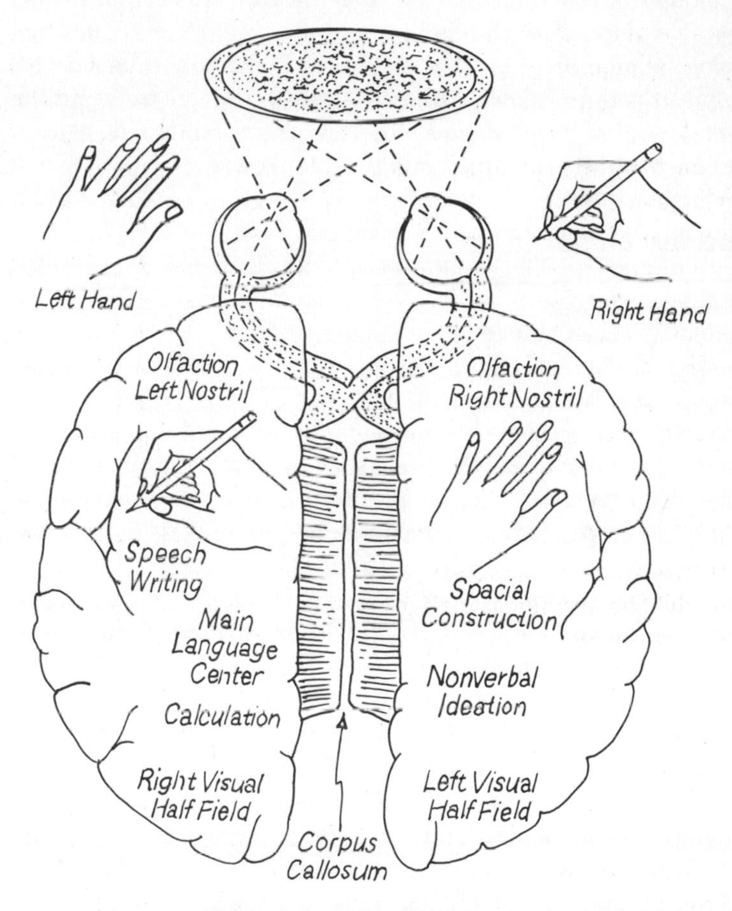

Figure 4-8 Corpus Callosum

The Right and Left Hemispheres of the Brain and Their Functions *Richard Restak, The Brain, The Last Frontier (New York: Warner, 1979), p. 190. Copyright © 1979 by Richard M. Restak. Reproduced by permission of Doubleday & Company, Inc.*

in processing and using information than had previously been thought.

At about the time of this medical research, Robert Ornstein, a psychologist, was studying the brain functioning of normal individuals. Using a mechanism that measures the alpha waves being emitted by each half of the brain, he attached plates that could register these electrical charges to different points on his subjects' heads. Knowing that alpha rhythms increase when the brain is at rest, he noted that, when the charge monitor went up, it meant that the part of the brain to which it was attached was not being used. Monitoring the charges while giving his subjects particular tasks to do, Ornstein was able to associate those activities with one or the other brain hemisphere. His research confirmed that verbal and numerical tasks apparently make heavy (if not exclusive) use of the left hemisphere while the right hemisphere rests, and that reproducing designs, grasping whole pictures, and recognizing faces occur mainly in the right hemisphere.[4]

Medical researchers are more cautious than the popularizers of the "split-brain" theory. They point out that hemisphericity, as they define it, should not be misinterpreted to mean that certain skills are lodged in one hemisphere and other skills in the other (although one can readily see how this conclusion could be drawn from the kinds of experiments Ornstein has done). Just because we know what we *cannot* do after the right hemisphere has been damaged does not mean that we know everything about what the right hemisphere can do *by itself.*

Thinking, as we know it, embraces acquisition of knowledge, memory, reflection, abstraction, and creativity. It relies on the healthy development of both hemispheres and, above all, on the unimpeded interaction of the two. Reading difficulties, according to this model of the brain, may result from weakness in the tissue between the two hemispheres rather than a poorly functioning left hemisphere. Such a weakness might prevent an otherwise normally functioning child from connecting the *picture* of a word, perceived through the right hemisphere, with the *sound* of that word that has been recorded in his auditory memory in the left hemisphere. The implications of this research for understanding and treating brain abnormalities are enormous. But what, if anything, does split-brain research contribute to our understanding of mathematics learning and possible sex differences in learning spatial skills?

Sex and the Brain: Lateralization

So far, nothing in split-brain research would especially account for male/female differences in spatial visualization. Humans are the only primates who experience handedness. But no more females than males are left-handed—a condition that in some but not all cases causes the hemispheres to reverse function. Most females, like most males, have speech lodged in the left hemisphere and the ability to perceive block designs lodged in the right.

But one sex difference has been noted, and that is the degree of specialization in the two hemispheres. Damage to one brain hemisphere sometimes has a lesser effect on adult women than a comparable injury has on men. Whether this is because a woman has more connecting fibers than a man between brain hemispheres, because a man's cortex is thicker than a woman's, or because hormones are at work again, no one knows.[5] Even more intriguing is the claim that there are sex differences in the *timing* of brain specialization.

We are born, it seems, with the potential to do all the known kinds of mental functioning in both halves of the brain. Children who experience a blow to the left side of the head severe enough to cause left-hemisphere brain damage often learn to speak using the right hemisphere, as long as the damage occurs before the age of seven or eight. But as we continue to grow, the adaptability of the brain's hemispheres comes to an end. Older children who receive severe brain damage to the left hemisphere may never recover speech or learn to read again, because by then their right hemisphere (the undamaged one) has lost its potential for verbal tasks. There is now some speculation that, mysteriously, this final specialization or brain lateralization takes place earlier in girls than in boys. This, it is argued, accounts for girls' earlier verbal aptitude; but the fixing of the hemispheres also causes a stagnation of their spatial capability at an earlier, perhaps premature, stage of development.

Biologist Anne Fausto-Sterling draws altogether different conclusions from the maturation thesis. The fact that the brain is "unfinished," so to speak, through early childhood means that environmental conditions—toys, games, athletics—can play an important role in spatial conditioning. Girls who are talked to more may develop verbal skills earlier and tend to rely on them even for spatial tasks. In any case, boys mature less quickly than girls, and this alone may account for the later brain specialization.[6]

But—and this remains the critical issue—is early (or late) brain lateralization good or bad? It appears that researchers cannot decide whether versatility is to be preferred to efficiency. But one finding is common, particularly in popular renderings of this complex subject: Whatever males have (or do) is better suited for math/spatial tasks than what females have (or do). Take the timing of brain lateralization. On the one hand, too-early specialization starves the hemispheres of their comfortable cross-competence. If you're having trouble doing a spatial problems on one side of the brain, and you're not yet lateralized, your brain can try the other. On the other hand, too-late specialization allows verbal tasks (in girls) to "spill over" to the right side of the brain, taking up valuable space (or time) that should be focused on spatial tasks.

The entire issue of brain lateralization is controversial and in flux. Julia Sherman, the psychologist who criticized the Embedded-Figure Test as a measure of analytic ability in females, also criticizes studies of bilateralization, and especially their conclusions about females.[7] She prefers another interpretation of the evidence: Earlier verbal development in females leads them to prefer a verbal approach. They learn by being told and later by reading, which is another form of being told. Thus, simply by preference, girls neglect the nonverbal approach to problem solving. They grow up not only to be more verbal than spatial but, Sherman speculates, altogether to prefer left-hemisphere thinking to right.

The relationship of spatial skills to mathematics is also far from clear-cut. Some researchers see spatial visualization as needed at all levels of math learning. Others find it high only among mathematicians who specialize in geometry or topology. Many people believe that the most important factor in predicting math aptitude is verbal facility, especially ability in reasoning and logic. Not all high-verbals do well in math, but apparently no one does well in math who is not also high in verbal skills. Even more confounding is the fact that there is not just one kind of spatial ability. Low-level spatial skills (as in figures 4-1 and 4-2) may not require any transformation of visual images. High-level spatial visualization (as in figures 4-3 and 4-4) is defined as ability to visualize three-dimensional configurations and to manipulate them mentally.

Even though not everyone agrees about the relationship between spatial visualization and math, it is important to deal with this issue. If it can be proved that something about female brain development makes it unlikely that females will enjoy and do well in math, there will be little interest in supporting ventures to cure their math avoidance. Since we know so little about how the brain develops, it seems to me we must deal with the issue in a different way.

Let us grant that the ability to do mental manipulations of visual images is probably useful in learning some parts of mathematics. We have observed that many females do not get the same exposure to spatial-visualization training in their play and their home environment. Thus, we must try to teach spatial skills to those who need them. Part of the effort being made at Mills College, a women's college in Oakland, California, to attract undergraduate women into higher math is based on these assumptions. Precalculus courses pay particular attention to familiarizing young women with graphing equations. The students learn to recognize the functions most frequently used in calculus and statistics problems before they study them in advanced courses.[8] Other math educators and specialists in spatial skills are looking for ways to teach these skills to children.

As athletic opportunities open up for girls and young women, sex differences in spatial skills may diminish even more. Studies of women who climb, do ball sports, and ski find that their spatial-relations abilities exceed those of average males. There are several possible reasons that athletes do better than average women and men. Moving their bodies through space helps them improve their sense of distance, and positioning their hands to catch or to return a ball may develop the ability to make quick mental projections of the ball's flight. The research is new, but it is suggestive, because, if spatial visualization is just another skill that can be learned, it is just another skill that can be taught.[9]

Improving Spatial Visualization

There is some evidence that participation in shop and mechanical-drawing classes also increases a student's ability to visualize objects in space. This, together with sports, might account for some male superiority over females. Until Title IX went into effect in the late 1970s, boys were required to take courses in shop and mechanical drawing in junior high school or high school, whereas girls were not allowed to take these courses and were required to take cooking and sewing instead. This forced specialization is now illegal, but there are still fewer girls voluntarily taking shop than boys. Taking shop decidedly improves spatial ability. Testing before and after the first year of engineering shows that the study of engineering can give the same results: Engineering students improve their spatial skills by studying engineering.

Another source of training in spatial visualization is boys' clubs. Boy Scouts traditionally learn to make and follow maps, use a compass, survey, and even appreciate architecture. They are given courses in "orienteering," learning to locate themselves in the woods or on a mountain with minimal mechanical aids.

Few similar options exist for girls. Girl Scouts may learn knot-tying (in some troops but by no means in all), but not orienteering. So we still wonder whether some boys are more skilled in this intellectual function because they are boys or because they are better trained.

Mapping Your Mind

Folding bed
Diaper
Butterfly net
Bird feather

List as many uses as you can for the objects named above. Use your imagination. The uses do not have to be conventional ones. There is no time limit. Hand in your paper when you are done.

This is a typical test of "divergent thinking." Students who enjoy an assignment like this one and who find endless possibilities in every object named are considered "divergent" thinkers. Divergent thinkers like open-ended tasks, tolerate ambiguity easily, like freedom to spend as much time as they want on a problem, and enjoy going off in all directions at once.

"Convergent thinking," on the other hand, is directed toward finding an answer or a unique solution. It is usually a process having an obvious beginning, a middle, and an end. "I've found it" means we can stop looking. "I've got it" means we can stop working.

It has been suggested in recent years that part of the reason for dislike of mathematics is that mathematics is perceived to require convergent thinking, the opposite of divergent. Those who do better at collecting information and working it down to a single solution are supposed to prefer math and science to history, literature, and even law. But the converse may not be true. In an important study, Lorelei Brush included the convergent/divergent dimension in her survey of attitudes toward mathematics. She found that really good science and math students do both convergent and divergent thinking.[10] Again, the issue seems to be ideological rather than actual: People who prefer divergent tasks associate math with a style they *think* is antithetical to their own.

Still, it would be interesting to find ways to teach mathematics to people who like shifting categories of thought, who find many answers to one question, and who produce ideas easily. One starting point might be with the order in which mathematical ideas have been conventionally taught. Teaching math as the ancients and moderns developed it, or as a discovery course, allowing students to find out rules and propositions for themselves, as Brush suggests, are two alternatives.*

According to one theory, we have to go through a four-stage cycle to learn anything at all. First we observe something through a concrete experience; then we reflect on what we have observed; we assimilate these observations into a theory, either one that we have thought about before or a new one; and, finally, we take some kind of action. Ideally, we should have all the abilities needed to observe, reflect, theorize, and act, and the most efficient learners among us probably do. But, given our various temperaments and the different learning experiences we have had as children and young adults, we may sharpen our abilities in one or two of these learning areas and underdevelop the others. For example, careful observation with attention to detail is in some senses the opposite of active decision-making and also differs from the ability to sort out information and create a theory. Reflection tends to inhibit action, and vice versa. In short, it is likely that we excel at and enjoy one kind of intellectual activity more than another.

One way to begin to get a picture of our individual learning style is to draw our own "cognitive profile." Sometimes called "learning-style inventories," there are instruments that have us rank ourselves according to some key words, like "discriminating," "tentative," "involved," "practical," "watching," "risk-tak-

*See chapters 6 and 7 for demonstrations of such an approach.

ing," and "future- or present-oriented." Our scores are then analyzed in terms of the four ways of knowing: concrete experience, reflective observation, analytic-conceptual, and active experimental. Another way to put the scores together helps us see whether we are "convergers" or "divergers."[11]

Another widely used typology is the Myers-Briggs "Indicator," a measure of personal preferences which, together, add up to a cognitive and/or learning style. "Sensing types," for example, take in information primarily through outside stimuli; "intuitive types" are more original in their thinking and have a "great drive for their own ideas and purposes." Overlaying these learning attributes is personality, which, the Myers-Briggs designers believe, also affects cognitive and learning styles. So, for example, extroverted sensing types are good at on-the-spot problem solving; introverted sensing types can apply themselves but only when they are already interested in a subject. A third dimension of the Myers-Briggs inventory measures "thinking" versus "feeling"; and a fourth measures "judgmental" (preference for living a planned and organized life) versus "perceptive" (preference for living a spontaneous and flexible life).*

According to those who employ the Myers-Briggs inventory to calibrate study skills and study habits, all of the composite types are capable of learning anything, including mathematics. But some are more attuned to the traditional way of teaching quantitative subjects than others.

The goal of all these "Indicators" is to gain an awareness of how we think and of differences in cognitive styles. Teachers of different subjects prefer certain learning styles, and very often our preference for a subject may be no more than a delight in the cognitive style we associate with it. Dislike of mathematics, from this perspective, may stem from dislike of the nondiscursive, didactic style of teaching. Conversely, disdain for history and social studies as taught in high school might be the result of having to learn facts devoid of theory.

*My thanks to psychologist William T. Harrison of Tucson, Arizona, who has made the Myers-Briggs instrument available to me.

All such superficial mappings of our minds must be treated cautiously, but it seems better to know something about how we learn before we try to alter our learning style or change the mathematics curriculum to fit our individual needs.

For one learner, it is better to start with theory and then go on to examples and applications. For another learner, the same information is best conveyed through discovery. Given some concrete materials to play with, you might learn more about dividing fractions and feel more secure than if your teacher began by explaining it to you. Some teachers do try to vary teaching techniques according to the various cognitive styles of their students. But if we are familiar with our own cognitive map, we can control our learning independent of the teacher's temptation to teach a concept as she would prefer to have learned it herself.

Adult learners might decide to start relearning mathematics with books on the history of math or the science of number, as I did. Others may choose to spend several weeks on puzzles to get their heads back into the rhythm of doing math. Still others will do quite satisfactorily with a review book that divides the subject into teachable bits, drills the material, and then tests for mastery.

Since our cognitive maps are forever being reshaped by what and how we learn, none of these preferences will remain fixed. It would be too much to ask any teaching institution to keep a running check of these changes. It is far wiser for us to monitor our thinking ourselves. But to do this we need to be self-confident about learning. We need to be able to say, "This approach does not suit me. Do you have another?" Or, "Can I read a book about calculus before I begin this class?" Or, "Give me an example of how this statistical procedure is applied in a real-life situation. Then perhaps I can figure it out for myself."

It would be a mistake to conclude that any self-administered test either is 100-percent accurate or describes a permanent, unchanging state. Like all self-tests, learning-style inventories cannot measure how we learn; we do not *learn* on the test, we only answer questions about ourselves. Rather, they indicate how we see ourselves as learners—our cognitive self-image. Even if at one stage of our lives we cannot learn from concrete experience, for example, it may be because of a childhood experience that, with patience, we can overcome. Mastering something new and difficult can affect our cognitive map in two ways: it can alter the way we think, and it can alter the way we think about how we think. If we see ourselves as impatient with detail, unable to recall numbers or amounts, incompetent at one sort of thinking or another, we may well perform or not perform just as we expect simply because we have a "cognitive self-image" as well as a cognitive style to contend with.

This does not mean that all our disabilities are self-imposed. Some experiences in learning mathematics are difficult for everyone, whatever her learning style or cognitive map.

Temperament

Sometime during the early years of the Math Clinic at Wesleyan, my colleague—a professional mathematician—and I decided to give ourselves a standard spatial-visualization test, the one used by the U.S. Army to screen recruits for certain training and jobs. The test consisted of sixty fold-the-box, embedded figures, and similar questions, to be completed in thirty minutes. Like obedient students, we set to work, with the timer clicking away, and after thirty minutes put down our pencils and scored our right and wrong answers. My colleague had completed only twenty-eight of the sixty problems. Working carefully and thoroughly, as mathematicians do, and not moving on to the next problem until he had checked and double-checked his answers, he only did twenty-eight problems, but he got all twenty-eight right. I, working at my usual fast pace, finished all sixty questions well under time, and got (it was sheer coincidence) twenty-eight right. What did we learn about our spatial-relations competency? Only this: I'd rather be finished than right!

There's a lesson in this anecdote for right- and wrongheadedness. Often our disabilities in doing a subject like mathematics have precious little to do with intelligence or brain organization. Patience and impatience, desire to see the whole before embarking on the parts of a problem, lack of attention to detail—these are all temperamental issues. We may not be able to change our temperament, but we can learn to exploit what we do best and mitigate the negative consequences of what we do less well.

Here's an example. When I attack a math or science problem, I use a "both-ends" strategy. I start with the question: What am I going to have to have at the end if I solve this problem correctly? Am I looking for the speed at which the vehicle is traveling, in which case my answer has to be in miles per hour? Or am I supposed to find out distance traveled, in which case the answer will be in miles? Once I decide what the answer should *look like,* then I go back to the question and try to figure out how to get there. Another student will proceed more methodically, in a forward direction, from the given information to the question. I prefer to go back and forth. This is not to say that methodological thinking is not important in mathematics; it is. But it may not be where every learner wants to start. It's the old woods-and-trees distinction.

In a year-long math-anxiety workshop, a woman learned that her impatience with math was linked to her cognitive self-image. Midway through the course, she told the instructor: "I used to think the smart people skipped steps. Wanting to be one of the smart ones, I was skipping steps. I finally learned not to do this. And I got more answers right."[12]

Math is only partly systematic. After you have figured out what to do, you will be able to proceed systematically through a learned procedure. But what do you do at the beginning, when you're not yet sure which formula to use or whether there is any formula at all? Alan Natapoff is a physicist at M.I.T. He notes that *groping* is something the human brain—even the best of them—does not do very well. He defines exploratory groping as seeking a particular object, solution, or approach in a very large field of possibilities. It is easy to find a pencil, he says, if we are told it is in one of two pockets. We will look systematically in one and then in the other. But if we are looking for lost keys that could be in any one of six pockets, or in a pocketbook, or in the car door, or on any one of four tables, or near the shelves, we grope unsystematically: into one pocket and out the other, looking in the same place many times over, getting more and more distraught as we continue our random search. According to Natapoff, math for the beginner, especially word-problem solving, requires just the kind of groping we do worst. No wonder inexperienced math and science students find the process upsetting and unrewarding.

Natapoff's system minimizes fruitless exploration. Lost keys are "placed," for the sake of having somewhere to begin, in one of two pockets. Not that they really are there, but at least the randomness of the search can be controlled by assuming for the moment that they are. Then, if they prove not to be there, the problem is reformulated: "Consider the car door and the door of the house." If this, too, turns out to be unfruitful, "Consider the end tables," and so on. The problem is not solved quickly this way. On the contrary, any experienced key-finder would have located the keys long before, using a tried repertoire of key-finding techniques. But at least the beginner can start somewhere and keep track of where he's been. Doing something minimizes the feelings of helplessness and fear.[13]

In fact, Natapoff's system applies quite well to problem solving at the higher levels. Encountering an unfamiliar problem, experienced mathematicians also must grope. Their "Let the basic unit equal passenger-late-minutes-per-mile" is one way to minimize groping. It is a form of "Let us assume that the lost keys are in one of two pockets." Then at least one can go to work.

Is There a Nonmathematical Mind?

> Math, along with related fields like chess and music, has a core of such pure intuition that a child of genius can display his powers quite early, not limited by experience as merely talented children are. Instead of soaking up knowledge faster than their peers, prodigies seem almost disconnected from experience.
>
> —T. Branch[14]

Is there a mathematical mind? Yes, of course. At some level of genius, connections are made that defy the normal learning pattern and eventually permit the individual to think thoughts no one has ever thought before. One reason for the early display of genius in mathematics is that mathematics can be developed as a system without any reference to reality or to experience. As a young child, Saul Kripke (about whom the above quotation was written) figured out for himself that $(a - b)(a + b) = a^2 - b^2$ just by playing with numbers and noting that sequences like $(7 - 5)$ multiplied by $(7 + 5)$ equaled the square of the larger number minus the square of the smaller one: $49 - 25$, or 24.

Still, this does not mean that the average person has a nonmathematical mind. For some, mathematical ideas may be self-explanatory; others need repetition and an opportunity to absorb more slowly. For one group of children, the numerals 1, 2, 3, and 4 may have more significance than the words "one," "two," "three," and "four" or the lengths – — —— and ———. From this perspective, mathematics anxiety may occur at the first encounter with another cognitive style or a jump in conceptualization for which the child or adolescent was neurologically or emotionally unprepared. But by the time a student has entered or graduated from college, she is not likely to suffer from any of the perceptual difficulties that may once have gotten in the way.

Some people, epitomized by the character in the following passage by Philip Roth, are so distracted by what Henry James called "felt life" that they cannot concentrate on the essential mathematical information as it is presented. They are so fascinated with detail, with the "people" part of issues, that abstracting the numbers and the ratios from a problem wrenches them from their real interests. Roth tells of Nathan, a sickly and feverish young boy, whose father tried to sharpen his mind by giving him arithmetic problems to solve. As Roth tells it, the father would announce a problem like this:

> " 'Marking Down,' " he would say, not unlike a recitation student announcing the title of a poem. "A clothing dealer, trying to

dispose of an overcoat cut in last year's style, marked it down from its original price of thirty dollars to twenty-four. Failing to make a sale, he reduced the price still further to nineteen dollars and twenty cents. Again he found no takers, so he tried another price reduction and this time sold it. . . . All right, Nathan; what was the selling price, if the last markdown was consistent with the others?" Or, " 'Making a Chain.' A lumberjack has six sections of chain, each consisting of four links. If the cost of cutting open a link—" and so on. The next day, while my mother whistled Gershwin and laundered my father's shirts, I would daydream in my bed about the clothing dealer and the lumberjack. To whom had the haberdasher finally sold the overcoat? Did the man who bought it realize it was cut in last year's style? If he wore it to a restaurant, would people laugh? And what did "last year's style" look like anyway? " 'Again he found no takers,' " I would say aloud, finding much to feel melancholy about in that idea. I still remember how charged for me was that word "takers." Could it have been the lumberjack with the six sections of chain who, in his rustic innocence, had bought the overcoat cut in last year's style? And why suddenly did he need an overcoat? Invited to a fancy ball? By whom? . . . My father . . . was disheartened to find me intrigued by fantastic and irrelevant details of geography and personality and intention instead of the simple beauty of the arithmetical solution. He did not think that was intelligent of me, and he was right.*[15]

The young Kripke was fascinated by the possibilities of number, the young Roth (for this tale must be autobiographical) just as much by the possibilities of personality. To call theirs "mathematical" and "nonmathematical" minds is to miss what they represent: two beacons in a continuum of human curiosity in search of meaning.

*I am indebted to Helen Vendler for reminding me of this passage.

Answers

Fig. 4-1, Turning-the-Square test: The correct answer is "B."
Fig. 4-2, Turning-the-Box-Test: The correct answer is "D."
Fig. 4-3, The Rotation Test, two-dimensional: The correct answer is "d."
Fig. 4-4, The Rotation Test, three-dimensional: The correct answer is "b."
Fig. 4-5, Finish-the-Square Test: Pieces 1 and 5 together finish the square.
Fig. 4-6, Fold-the-Box Test: Box 1 is correct.
Fig. 4-7, Embedded-Figure Test, Figure "a" is located in box 1; "b" in boxes 2 and 5; "c" is not in any box; "d" is in boxes 4 and 6; "e" in box 3.

Notes

1. Anne Fausto-Sterling elaborates on this point in *Myths of Gender: Biological Theories About Men and Women* (New York: Basic Books, 1985), pp. 33–36.
2. Doreen Kimura, "Sex Differences in the Brain," *Scientific American*, Sept. 1992., pp. 123–25.
3. Roger Sperry, "Some Effects of Disconnecting the Cerebral Hemispheres," *Science*, vol. 217 (1982), pp. 1223–26, as quoted in Fausto-Sterling, *Myths of Gender*, p. 47.
4. Robert E. Ornstein, "Right and Left Thinking," *Psychology Today*, May 1973, pp. 87 ff.
5. Kimura, "Sex Differences," pp. 123–24.
6. Fausto-Sterling, *Myths of Gender*, p. 48.
7. Julia Sherman, "Effects of Biological Factors on Sex-related Differences in Mathematics Achievement," paper prepared for the National Institute of Education, Fall 1976. See also Julia Sherman, *Sex-related Cognitive Differences: An Essay on Theory and Evidence* (New York: Charles C. Thomas, 1978).
8. Lenore Blum, professor of mathematics at Mills, has been in charge of this program.
9. Marilyn Jones, "Perceptual Studies, Perceptual Characteristics, and Athletic Performance," in H. T. A. Whiting, ed., *Readings in Sport Psychology* (Lafayette, Ind.: Balt Publishing, 1972).
10. Lorelei Brush, "Mathematics Anxiety in College Students," *Journal of Counseling Psychology*, vol. 19 (1978), p. 551.
11. *Learning Style Inventory: Self-scoring Test and Interpretation Booklet*

(Boston: McBer and Company, no date). Available from McBer and Company, 137 Newbury Street, Boston, Mass. 02116.

12. Taken from the film *Math Anxiety: We Beat It, So Can You,* 1980, available from Educational Development Corporation, Newton, Mass.
13. Alan Natapoff, personal communication.
14. T. Branch, "Saul Kripke: New Frontiers in American Philosophy," *New York Times Magazine,* Aug. 14, 1977.
15. Philip Roth, *My Life as a Man* (New York: Holt, Rinehart and Winston, 1974), pp. 36–37.

5 Word-Problem Solving: The Heart of the Matter

de Weese: How did you know the problem was in the switch?
Phaedrus: Because it worked intermittently when I jiggled the switch.
de Weese: Well, couldn't it jiggle the wire?
Phaedrus: No.
de Weese: How do you know all that?
Phaedrus: It's obvious.
de Weese: Well, then, why didn't I see it?
Phaedrus: You have to have some familiarity.
de Weese: Then it's not obvious, is it?

—Robert M. Pirsig, *Zen and the Art of Motorcycle Maintenance*

Given what we already know about the causes of math anxiety, I wish I had the power to change certain things about the teaching of mathematics in the United States. Among them, I would abolish timed tests, by decree if necessary. I would let children use an abacus or calculator to diminish the nervousness that comes from fear of forgetting. I would teach spatial visualization in school, especially to little girls and women; like Pascal's bet on the existence of God, even if it turned out not to matter, we would have lost nothing by teaching it. And, not least,

I would figure out some way to help people conquer their fear and disability in solving word problems.

Word problems, sometimes called "story problems" or "statement problems," are, in my opinion, at the heart of math anxiety. They appear throughout the elementary curriculum and are the first reasoning problems of any sort that children are given. Since such problems prefigure science problems, students need learning strategies to solve word problems if they are not to grow up avoiding science and math.

More than any other aspect of elementary arithmetic, except perhaps fractions, word problems cause panic among the math-anxious. We have seen already how the math-anxious refuse even to attempt the Tire Problem, and how word problems depress and defeat them. Their attitude, not their math ability, gets in the way.

Lynn Fox asks students enrolled in her math-enrichment program what they do when faced with a difficult math problem. Do you stay with it until you have solved it? Do you leave it and return to it later, refreshed? Do you go to someone for help? Or do you forget it? We can classify people by their answers to this question, not because the answers reveal their mathematical aptitude but because they tell us much about their expectations of themselves.

Most difficult problems are not immediately obvious to anyone (even to Phaedrus). There is a good reason why even many capable mathematicians do not like to do math in public. For a period of indeterminate length, one flounders. How well one sticks with a problem through this floundering may well be a function of one's tolerance for floundering in general, or of how well one flounders in math. It would help us to know what mathematicians do when they are floundering. Do they enjoy it? Do they busy themselves with some kind of operation, something between a doodle and a stab at the problem? Do they sketch it? How can one flounder constructively?

At its most destructive, floundering creates a panicky search for some formula that will liberate one magically from the dilemma. Even when confronted by a puzzle without numbers, or by a question that can legitimately be answered in terms of unknowns (like the following), people will seek a formula.

> A man earns a monthly salary. At Christmastime, he is given an extra month's salary as a bonus. How much does he earn in that year?

Since no numbers are given, it will be impossible to answer this question in dollars. But the information asked for can be expressed in the terms that are given. There is an unknown: the monthly salary. The employee is given one more of those than there are months in the year. Hence he will earn 13 times his monthly salary, or $13x$; no numbers, no formula, only some degree of recasting of the problem.

Another reasoning problem goes like this:

> A tennis tournament has 125 contestants. On losing a match, a contestant drops out of the tournament. The winner must win all the matches he or she plays. How many matches are needed to complete the tournament?

Here again there is no formula. One might be tempted to draw some kind of tree diagram starting with all the players, playing them off one against another. But if we turn the question around and ask how many people must be eliminated to produce one winner, the answer is easier to find: 124 people will each lose exactly one match. Hence 124 matches will produce a winner.

These are tricky questions in a sense, because we expect to have to do some computation and it turns out we do not.* But, as the psychiatrist Michael Nelson points out, most problem-solving mistakes are neither computational errors nor errors in logic. Rather, they are psychological in origin. The "I can't" syndrome seems to be particularly disruptive in doing word problems.

How can we learn to handle word problems? One insight, provided long ago by Robert Davis at the University of Illinois, is that,

*Tricky questions are fair; trick questions are not.

when people come up with wrong answers to word problems, they may have the right answer to another question. Therefore, a key to treating fear of word problems is to discover the question the student was answering by analyzing the error he made. We can also do that kind of analysis for ourselves.

The Tire Problem Revisited

Consider the Tire Problem again:

> A car goes 20,000 miles on a long trip. To save wear, the 5 tires are rotated regularly. How many miles will each tire have gone by the end of the trip?*

Many people find the Tire Problem difficult, but not because of the mathematical reasoning or the calculations. The notion of rotating the tires is foreign to them. One woman comments: "First I had to imagine what was being talked about, which made me anxious because I wasn't sure I was right. Perhaps there were some rules about rotating tires that I didn't know." Another writes about how a group of adult women dealt with the problem: "Very few people used any common sense with it at all. Some supposed it meant that the car had five wheels. Others thought you had to figure out the rotation or envision some kind of continuously changing set of tires." This is not as unusual as it sounds. Many people get bogged down looking for a mental picture to associate with the problem.

No amount of contempt or condescension can eliminate difficulties such as these. But probing wrong answers can sometimes bring these false notions to the surface, which documents how irrelevant to the solution these issues really are.

Two of the most common wrong answers to the Tire Problem

*Treatment of the Tire Problem is derived partially from systems developed by Susan Auslander and reported in her "Teaching Word Problems in the College Math Class," unpublished paper, 1977.

are 4,000 miles and 15,000 miles. The first answer, 4,000 miles, was arrived at by dividing the number of miles traveled by the number of tires, 20,000 divided by 5. This answer is not right, but it can be described as the answer to another interesting and important (if preliminary) question: How many miles will each tire spend in the trunk during the trip? Once that question is correctly formulated, it is only one more step to the right question and from it to the right answer—20,000 miles, the total mileage, minus 4,000 miles, the number of miles any one tire will have been in the trunk, is 16,000 miles, the number of miles any one tire will have been on the road.

The second answer, 15,000 miles, comes from a similar realization that each tire will be in the trunk some portion of the time, in this case (incorrectly) ¼ of the time. Thus, it was assumed incorrectly that the tire will be on the car ¾ of the time, and ¾ times 20,000 equals 15,000. This answer is not correct because each tire is off ⅕ of the time, not ¼. But the approach was intelligent.

Now consider several ways to get the correct answer, 16,000 miles. One way is to notice that 4 of the 5 tires are being used at any given time and to take ⅘ of 20,000 and get 16,000 miles. But this is not the only way. We can think of time in trunk as ⅕ of the time, leaving ⅘ of the time for time on wheels. A third is to reason: If the car goes 20,000 miles, then the tires together drive 80,000 tire-miles. If each of the 5 tires is used equally, as we have been told, then each tire will go 80,000 tire-miles divided by 5 tires, or 16,000 miles. Note how useful the unit tire-miles turns out to be.

There are sound pedagogical reasons for being flexible with one's students or with oneself. One person's tidiness is another person's mess. One person likes to go from knowns to unknowns, another from unknowns to knowns. One person will start by sorting out the detail, another by getting the whole picture, and others by achieving different combinations of the two. No matter. The key is finding a way that keeps one at the problem, because the people who lose are not the ones who are wrong but the ones who give up.

The Painting-the-Room Problem

> If I can paint a room in 4 hours and my friend can paint the same room in 2 hours, how long will it take us to paint the room together?

The first wrong answer that many people come up with is 3 hours. This is plainly the result of adding the two amounts of time and dividing by 2: 4 hours plus 2 hours equals 6 hours, and 6 divided by 2 equals 3. Logically, this answer does not make sense, though it is computationally correct. If it took both of us 3 hours to paint the room, my friend, who could paint the room alone in 2 hours, would not need my help. Even though the answer is absurd, however, it does show us that the correct answer is going to be less than 2, and this puts us somewhat ahead.

One constructive approach to the problem is to draw a picture of the room (figure 5-1).*

On this model we see how much is painted in some common unit of time by both of us. We choose as a common unit 1 hour. My friend can paint ½ the room in 1 hour; I can only paint ¼ of the room in that time. Thus, at the end of 1 hour painting together, we have covered ¾ of the room (or 3 out of 4 walls). There is ¼ left to do. If ¾ of the room gets done in 1 hour, then the remaining ¼ will get done in ⅓ of an hour. Hence, the whole room gets painted in 1⅓ hours. The appeal of this approach is that there is no algebra, no ready-made formula, only a step-by-step working out of the situation. Note also the options: We can consider the room as a whole and think about ½ the room, ¾ of the room, or ¼ of the room. Or we can think about it wall by wall (see figure 5-1 again). We can compute the hours and parts of hours, or we can turn the hours into minutes and work with that. We can work out the entire time or choose some unit, 1 hour, and work from there: many ways, many pictures, many solutions.

*Treatment of the Painting-the-Room Problem is also derived from Susan Auslander's "Teaching Word Problems in the College Math Class." Note that it is assumed, though not stated, that painting the room means painting the four walls, not the ceiling, too.

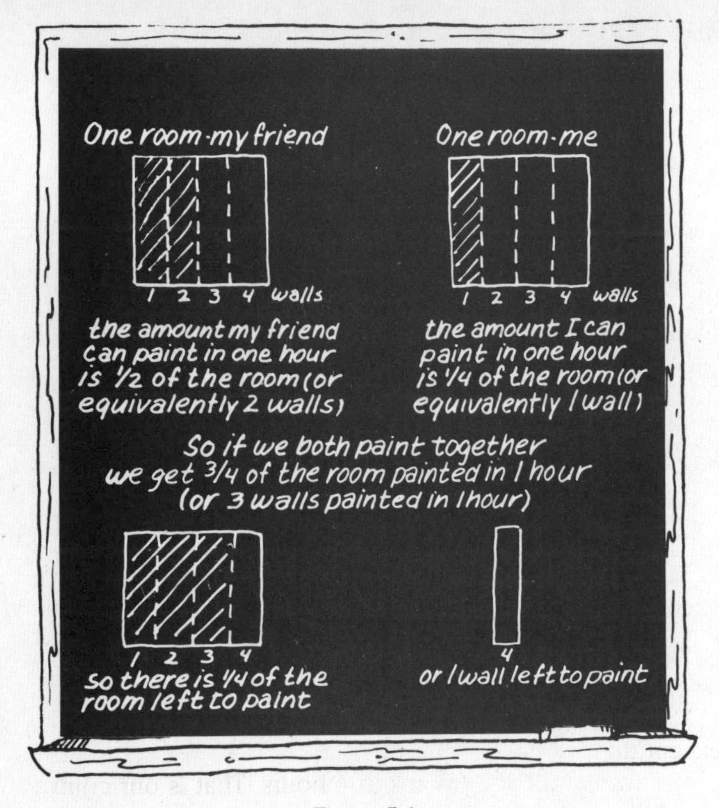

Figure 5-1

This approach to the Painting-the-Room Problem courtesy of Susan Auslander.

A less visual way to solve this problem is to say,

Let t = time in hours we spend painting the room together.

This is a way to give the final answer a designation, t, before we know what its value is. The process we follow is algebraic.*

The trouble with an algebra approach for people who are not comfortable with it is that, once we have designated the answer,

*Let t = the time it takes us to paint the room together. In t hours, I paint $\frac{t}{4}$ of the room and my friend paints $\frac{t}{2}$ of the room. $\frac{t}{4} + \frac{t}{2}$ must equal 1 room.

in this case *t*, we feel as if we are leaving the situation, and this makes us insecure. How do we know whether we are on the right track when we are dealing with fractions and "*t*"s? One way to feel more secure about the problem, I think, is to stay closer to the information given, not to abstract it quite so soon.

We know that 1 room can be painted in 2 hours by 1 person, and that that same room can be painted in 4 hours by another. Adding the hours will get us nowhere, because the total time is going to be less than if both were doing the room separately. Perhaps adding the *rates* at which the two people work is the direction in which we ought to go. But how do we add rates? This is a good question and, as we will see, it takes us to much more sophisticated mathematics.

Let us begin by calling the rates "fractions," because fractions designate ratios as well as parts of a whole. (See chapter 6 for more on the different meanings of fractions.) My rate is $1/4$ (1 room in 4 hours). It will always be $1/4$ whether I paint 2 rooms in 8 hours ($2/8$), 3 rooms in 12 hours ($3/12$), or 50 rooms in 200 hours ($50/200$). My friend's rate is $1/2$, as it, too, will always be whether she paints 3 rooms in 6 hours ($3/6$) or 50 rooms in 100 hours ($50/100$).

Now, adding the rates, we get $1/2 + 1/4 = 3/4$. So far so good, except that we have forgotten by now what $3/4$ means—that together we can paint 3 rooms in 4 hours. That is our combined rate. But since the question was not how many rooms can we paint together in 4 hours, but how long it will take us to paint 1 room together, we have to change that rate from 3 rooms in 4 hours to 1 room in so many hours. The way we do that is the nub of the operation.

Well, we can say, "Three rooms in 4 hours changed to 1 room in so many hours," several times, until we get an idea of how to proceed. (Not a bad way.) Or we can express this, too, as fractions:

$$3/4 = 1/?$$

If at this point we notice that the numerator (3) has been divided by 3 to get the new numerator (1), then, remembering the rules about fractions, we might consider dividing the denominator (4) by 3 to get the new denominator, 4/3. This produces the right answer (1⅓ hours), but it is not the only way to get the answer. Another way, once we have found the combined rate (3/4), is to divide it into one room:

$$\frac{1}{\frac{3}{4}} = \frac{4}{3}$$

and arrive at 1⅓ hours, too.

This problem seems difficult to do because it *is* difficult. Adding or comparing rates is not a simple matter. The calculations may be simple, but the principle behind them is quite advanced. So, if the Painting-the-Room Problem gives us trouble, it is not because we are dumb but because we are smart enough to intuit that there are complex ideas just beneath the surface.

Currents and Speeds

Another rate problem produces a similar difficulty.

> A ship goes in one direction, west to Hawaii, at 20 nautical miles per hour and, because of the wind, makes the return trip at 30 nautical miles per hour. What is its average speed?

The first time they hear this problem, many people will react to it as to the Painting-the-Room Problem and average 20 nautical miles and 30 nautical miles to arrive at an "average speed" of 25 nautical miles per hour. The mistake here is to assume that both trips took the same number of hours. In fact, both trips covered the same distance, but not in the same time. Take any distance, just to check this out. For a common distance of 600 nautical miles, going one way (600 ÷ 20) would have taken 30 hours; going the other way (600 ÷ 30) would have taken 20 hours. Thus, the round trip would take 50 hours, and 50 hours divided into

1200 nautical miles gives an average speed of 24 knots.* Again we are trying to compare rates, and the way we compute average rates, as we would have learned if we had gotten to statistics, is more complicated.†

Although the Ship Problem is not a calculus problem, comparisons of rates of change are the kinds of problems found in mechanics and dynamics, out of which calculus developed. In all these problems, the calculations are simple once we have formulated the problem correctly, but formulating the problem is not at all easy.

Another example is the problem of upstream travel against downstream currents.

> If a boat travels in still water at 6 miles per hour and the downstream current is 3 miles per hour, how long will it take the boat to go 10 miles upstream?

The computation is deceptively simple. Actually, we just subtract to get the rate per hour ($6 - 3 = 3$) and divide the total miles (10) by the rate per hour (3).‡ But the person who likes to think through and understand such a problem may have a harder time, because it is hard to figure out how fast or slow the boat is going

*A neat formulation for this is:

$$\frac{D}{20} + \frac{D}{30} = 2 \text{ (Trips)} \qquad 30D + 20D = 1200$$

$$50D = 1200$$

$$D = 24$$

†Averages are computed, as most of us know, by adding up the elements and dividing by the number of items added. Average rates are computed by taking the reciprocals of the numbers, $\frac{1}{20}$ and $\frac{1}{30}$ in this case, adding these, and dividing them *into* the number of items, in this case:

$$\frac{2}{\frac{1}{20} + \frac{1}{30}} = \frac{2}{\frac{3}{60} + \frac{2}{60}} = \frac{2}{\frac{5}{60}} = \frac{2}{\frac{1}{12}} = 2 \times \frac{12}{1} = 24$$

‡See answers for this chapter, p. 164.

at any one instant. Does it go 6 miles an hour and then get set back? Or does it somehow go backward and forward at the same time? It is not easy to form a mental picture of what is going on. Movement is very difficult to visualize.

Nonnumerical problems are sometimes even more difficult to handle than the ones that can be reduced to numbers. Take the problem:

> If we were to lay dominoes on a checkerboard so that each domino covered 2 adjacent squares, how many dominoes would we need?

Many people will simply go blank when confronted by such a problem. For one thing, they don't have a clue to how many squares, black or white, there are on a checkerboard. They cannot even begin to think about the problem. If we are told that there are 32 black squares and 32 white squares and we know that each domino is to cover 2 adjacent squares, we can figure out that there will be 32 dominoes on the board. Mathematics helps here not by providing any formulas or systems of calculation, but by suggesting a way to organize knowns and unknowns. But it only helps if we are on the right wavelength to begin with.

Cultivating Intuition

Obviously, problem solving is not really a matter of making logical deductions from memorized formulas, but an exercise in imagination. Cognition has sometimes been defined as "seeing relationships." It is that and more: It is fantasizing about relationships, trying out imagined ones until one is found that fits the situation. The use of tire-miles in the Tire Problem is a good example of imagination at work. The relationship that links all the tires and all the miles driven has to be construed out of nothing and provides an interesting way to get control of the problem.

No one can consciously control the image-making part of the brain, the faculty of intuition or insight. Beginners and people who have not done math problems for a long time come to believe that because they do not have instant flashes of insight they have no intuition. Since it is always easier for the teacher, the tutor, or the text to provide one image for the learner to apply than to wait for the learner to develop her own, some people never even find out that they can invent images for themselves.

Mathematical intuition is often mistaken for that mystical "mathematical mind" so many people are persuaded they do not have. Actually, intuition can be developed like any other skill. It responds to exposure to math and to other related experiences. Phaedrus' "intuition" about the faulty switch, quoted at the beginning of this chapter, was nurtured in long hours of fixing electrical connections. What was obvious to him was not at all obvious to someone who had not had the same experience.

An example of seeming linguistic "intuition" illustrates this point. The beginning German student with little background in other foreign languages and a not very sophisticated knowledge of English will have to memorize the distinction between the words *abreisen,* to travel away from, and *anreisen,* to travel to. Other students, just as new to German but already familiar with Latin, may intuit the difference without much effort, but that is because they already know that in Latin *ab* usually means "away" or "from" and *an* means "to" (abnormal, announce, annex, etc.).

The inexperienced beginner, to take another example from German, may have trouble translating the phrase *Im Laufe des Tages* ("in the course of the day") even when he knows the meanings of the words *Tag* (day) and *laufen* (run). A student who is acquainted with French or Latin, knows a little Italian from home, or has a more sophisticated knowledge of English, however, will probably guess correctly that in this phrase *Laufe* does not mean "running" but "course."* Just the same, what seems like a better language facility is only more experience with language.

*As in the French *courir,* to run, or *au courant* for being on top of things, etc.; or the English "current" for running water.

Mathematical intuition at the levels we are discussing may not be so different from other kinds of intuition. The more experience one has in solving problems, and the more varied one's mathematical repertoire, the more facile and intuitive one will appear. To cultivate our mathematical intuition, then, we must collect and keep fresh lots of pieces of information and many kinds of strategies. Then, when we need to, we can quickly search among a rich store of ideas for those that will help solve the problem at hand.

Guess-and-Check

"Don't guess!" our teachers told us, convinced that they were protecting us from carelessness in our work. But, in fact, educated guesses are the stuff of mathematics. At the highest level of mathematical-problem solving, professionals very often guess (they call it "hypothesizing") what might be the solution or the logical pattern that will unlock an answer or a proof. What makes guessing okay in advanced mathematics is this truth: The process of checking one's guess very often mimics the algorithm or formula by which the problem will eventually be solved.

Nancy Angle is a mathematics educator who has taught many levels of mathematics for over thirty years. She currently works with underprepared students and returning adults as well as undergraduates at Cerritos College in California. She encourages her students to use systematic trial-and-error when they are learning to solve word problems, especially in beginning algebra. Contrary to what you may think and what teachers may tell you, you can *begin* solving a word problem without first writing an equation. Not all problems are amenable to guess-and-check, she admits. But many are, and this approach will introduce you to mathematical thinking. Besides, it gives you something to do while chasing an elusive formula.

Take the following problem:

A cyclist bikes west at 12 miles per hour. A walker leaves from the same point at the same time and walks east at 3 miles per hour. In how many hours will they be 45 miles apart?

Let's assume you're stumped by the problem and don't think to let x equal the number of hours that it will take for the 2 travelers to be 45 miles apart. If you start, instead, with an educated guess, say, that it takes 5 hours, and do the calculations (see the first line in the table below), you will find the total mileage too high. This means that you will want to lower your guess to, say, 4 hours. Doing the calculations (see line 2 below), you will find the total distance apart still too high. As you guess your way to the right answer (see line 3 below), you are showing yourself the method for writing the algebraic equation.

YOUR GUESS	CYCLIST	WALKER	DISTANCE APART
5 hours	60 miles	15 miles	75 miles
4 hours	48 miles	12 miles	60 miles
3 hours	36 miles	9 miles	45 miles

Notice how you checked your answer. You took the guess number, multiplied it by 12 (the cyclist's miles per hour), multiplied the same guess number by 3 (the walker's rate), and added the two products together. Then you checked to see if the sum was equal to 45, the desired distance apart. In more symbolic shorthand, you might say you had constructed a formula (just another way of saying "an equation") that looked like this:

$$12 \times \text{guess number} + 3 \times \text{same guess number} = 45$$

Now all you have to do is to replace the "guess number" with a variable such as x, and you have an equation which the rules of algebra will permit you to solve.

$$12x + 3x = 45$$

$$15x = 45$$

$$x = 3$$

This answer is a nice neat number, but that is often not the case. Suppose the question had been, "When are the cyclist and the walker 50 miles apart?" From your guesses, you would know that the answer is somewhere between 3 and 4 hours, but is not a whole number of hours. However, you can now write the same equation with the number 50 and solve it to find an answer you probably wouldn't have guessed—3⅓ hours.

Before we leave this rather typical textbook problem, Nancy Angle suggests that you notice a few things. First, it was necessary to make some assumptions, which in the real world may not make sense. The textbook authors expect you to assume that (1) both travelers continue at the same pace, without stopping or resting, for the entire time; 2) the road on which they travel is a straight line which goes due east and west for the entire 45-mile distance. They also expect you to know the basic relationship that "distance equals rate times time" or d = rt, and to recognize that you must *add* the two distances from their starting points to find the distance apart. Here again, words make trouble. You might have thought, from the word "apart," that you were looking for a *difference,* and difference in mathematics usually implies subtraction.

Every word problem, she notes, must make some assumptions. If none were made, every problem would be pages long. For instance, in the classic problem of shooting an arrow at a falling apple, in which you are asked to figure out how long you should let the apple fall before shooting, the problem never tells you the arrow has to be aimed in the direction of the apple. Or, in the Painting-the-Room Problem (see page 137), any teenager knows that, if Tom and Suzy are working together, they won't work as efficiently (it's too tempting to socialize) than if they were working alone. Part of the art of doing word problems is recognizing what missing information you are expected to supply—and this gets easier with practice.

The trial-and-error method is helpful in solving many types of typical algebra word problems. However, you want do more than just guess. You want to learn to use algebra to find the answer. To do this, you must work somewhat systematically. Making a chart can be helpful. Here's another problem from Nancy Angle's collection. Try it.

> The sum of 3 consecutive odd numbers is 417. Find the numbers.

Guess-and-check can be tedious, and that's where algebra comes in. Consider another of Angle's problems: One side of a rectangle is 11 feet longer than the other. Find the dimensions of the rectangle if its perimeter is 96 feet. (They assume you know the formula for perimeter, 2 (side 1 + side 2)). Using trial and error, you would come up with a table like this:

SHORTER SIDE	LONGER SIDE	PERIMETER
20 ft.	31 ft.	2 (20 + 31) = 102 too big
15 ft.	26 ft.	2 (15 + 26) = 82 too small

Whenever you've figured out how to express the procedure you're using to come up with a solution that you then check, you can stop guessing and use a variable. Your next line in the table, then, would look like this:

s	$s + 11$	$2[s + (s + 11)]$

Now you are ready to write the equation, $2[s + (s + 11)] = 96$, and solve it to find the solution $S = 18\frac{1}{2}$.*

Learning to solve word problems is difficult, but, much as you want to avoid them, they just will not go away. Though using guess-and-check may take more time, it has at least two advantages: You can get correct answers, which is better than no answers or a wrong answer. If you pay attention to how you guessed your way to the correct answer, you will teach yourself, naturally

*Remember that the solution to the algebraic equation may not *always* provide the answer to the original question asked. In this case, the rectangle's *dimensions* are asked for, so the answer is $18\frac{1}{2}$ ft. (s) by $29\frac{1}{2}$ ft. (s + 11). It's always a good idea to reread the problem after you've solved it, to make certain that the solution you're about to write down matches what is asked for.

and less painfully, how to solve the same problems, using the more efficient method of writing algebraic equations.

Letter Division Problems

In my household, we were fortunate to have a parent who was not just a puzzle solver, but a puzzle maker. Looking back over the hours we children spent solving "Letter Division" problems of my father's design, I can see clearly now how much we learned in the process about *guess-and-check, persistence, keeping a record of what we have found out*—all of which adds up to feeling *comfortable* and employing *creativity* in problem solving (also affording us much informal practice we got at home in mastering our arithmetic facts).

Letter Division is just that: a long-division problem employing letters in place of numbers. The task is to figure out which letter represents which digit, such that the division problem is correct. Take this one, for example, a play on the puzzle itself, which has to contain "only ten digits" to work:

```
1                      T E N
                 -------------
2   O N L Y  ) D I G I T S
3              L E G S
               -------
4              N E S I T
5              N T Y Y Y
               ---------
6                  Y S G S
7                  O N L Y
                   -------
8                    L L Y
```

Obviously, this puzzle requires some knowledge of how to do long division, so I asked Mary Ellen Hunt, a Letter Division whiz, to provide a review. Thanks to the calculator, long division is rapidly becoming a lost art. But, Mary Ellen tells us, it remains an important exercise in putting a variety of skills to work. From the time of the Greeks, long division was thought to be hard precisely because it requires a certain mastery over all the other arithmetic operations—addition, subtraction, multiplication, and division. But that's what makes it useful, and fun, too. Though it's a serious challenge to our problem-solving skills, it doesn't require any knowledge of higher mathematics. In the next example, we see a long-division problem in numerals. It might have a real-world connotation—e.g., how many bags of grain can be fairly *distributed* among 4,185 starving Ethiopian villages if we have 967,620 bags of grain to begin with? Translated into math, the question asks: How many times will 4,185 "go into" 967,620?

With numbers this large, it would be impossible to solve the problem in our heads. That's why the *long* system for solving long-division problems was invented. First, as we see from the way the problem is set up, Mary Ellen makes a drawing: She places the number to be divided up (the *dividend*) under an open-ended rectangle. That permits her to place the number that must divide it (the *divisor*) outside the block, and to write the answer, digit by digit (the *quotient*), on the top bar. The system is thought to have derived from Hindu mathematicians. The Romans called it the "galley method," because the shape of the columns reminded them of galley ships.[1] The advantage, then and now, was that we can write the quotient digits in the right decimal places above the dividend digits.

Solving a long division problem involves a series of guess-and-checks, trials and approximations of an answer.

```
             2 3 1
4 1 8 5 ) 9 6 7 6 2 0
          8 3 7 0
          -------
          1 3 0 6 2
          1 2 5 5 5
          ---------
            5 0 7 0
            4 1 8 5
            -------
              8 8 5
```

Answer: 231 bags of grain per village. (The remainder is of little consequence.)

```
1                 T E N
2  O N L Y ) D I G I T S
3            L E G S
4            N E S I T
5            N T Y Y Y
6              Y S G S
7              O N L Y
8                L L Y
```

Now let's take a look at the Letter Division puzzle shown above. A good first step in solving any puzzle is to look for the obvious or easy parts—a kind of point of entry. For example, letters that might equal zero or one. One tends to show up in the multiplication steps since any number multiplied by one equals itself. Zero shows up in the subtraction steps. In this case, I − S = 0, so S = 0. One might also notice that on line 7 the letters O, N, L, Y are exactly the same as the divisor. If N × ONLY = ONLY, we can deduce that N = 1 (one), so now we have two places to fill in: S = 0 and N = 1.

Step	one
9	
8	
7	
6	
5	
4	
3	
2	
1	N
0	S

In the interests of keeping track of our work, we could make a grid like the one to the left, and fill in the number equivalents as we figure them out. Another helpful diagram would be a long-division "blank," which can be filled in as we go along, so that we can look for patterns.

```
1                 _ _ 1
2  _ 1 _ _ ) _ _ _ _ _ 0
3            _ _ _ 0
4            1 _ 0 _ _
5            1 _ _ _ _
6              _ 0 _ 0
7              _ 1 _ _
8                _ _ _
```

```
1                    T E N
2   O N L Y ) D I G I T S
3             L E G S
4             N E S I T
5             N T Y Y Y
6                 Y S G S
7                 O N L Y
8                   L L Y
```

Back to the puzzle.

Notice that T × ONLY = LEGS. This may be a little more complicated, because in multiplication (and subtraction and division) there may be "carryovers" to the next place.

As an aside, here's an example of what I mean. Take the problem of 4 × 78. To the left, we see it in more conventional form. To multiply 4 × 78, we begin by multiplying the simpler 4 × 8, which is 32. We write the 2 down and then carry the 3 over to the next place. Now we multiply 4 × 7, which is 28, and add on the 3 from the first part. Thus, we get 31. Writing that down next to the 2, we see that 4 × 78 is 312.

```
  7 8
×   4
? ? ?
```

$4 \times 8 = 32$

```
  7 8
×   4
? ? 2
```

$4 \times 7 = 28$

$28 + 3 = 31$

```
  7 8
×   4
3 1 2
```

Because of the "carryovers," our answer doesn't really look a lot like either the 32 or the 28 that were the solutions to our simple multiplications (4 × 8 and 4 × 7). It is this "carryover" which we'll need to watch out for when solving parts of the Letter Division puzzle.

```
1             T E N
2  O N L Y ) D I G I T S
3            L E G S
4            N E S I T
5            N T Y Y Y
6                Y S G S
7                O N L Y
8                  L L Y
```

We can't rule out the possibility that this is the case, but since T × Y = S or zero, LEGS must be a number ending in zero (10, 20, 30, etc.) The only way arithmetically to get a product ending in zero is by multiplying any number by zero or an even number by 5. Since S = 0, we can deduce that either T or Y must be 5 and that the other number must be even.

```
1                 _ _ 1
2  _ 1 _ _ ) _ _ _ _ _ 0
3            _ _ _ 0
4            1 _ 0 _ _
5            1 _ _ _ _
6                _ 0 _ 0
7                _ 1 _ _
8                  _ _ _
```

In lines 1, 2, and 5, we note that E × Y = Y (or some number ending in Y), and also, in lines 6 and 7, that S − Y = Y, or 0 − Y = Y. The zero implies that we will have to borrow from the tens column to make this 10 − Y = Y. Thus, we can see that Y must be equal to 5. We should also keep in mind that the G in line 6 was borrowed from. It may help us later.

```
1                 T E N
2  O N L Y ) D I G I T S
3            L E G S
4            N E S I T
5            N T Y Y Y
6                Y S G S
7                O N L Y
8                  L L Y
```

Now we notice that, in lines 4, 5, and 6, I − Y = S or I − 5 = 0. (A few correct numbers can open up a lot of doors.) Since I cannot be 5 (Y = 5), a good guess is that the number in the ones column needed to borrow from the tens column in the subtraction process, in order to make itself large enough. That means that I must have been a 6, was borrowed from, and ended life as a 5.

Steps

	one	two
9		
8		
7		
6		I
5		Y
4		
3		
2		
1	N	
0	S	

Going back to the idea of multiplying the answer times the divisor, since Y = 5. E must be an odd number. (Remember, even numbers times 5 produce numbers ending in zero.) E × Y = Y or some number ending in Y. E can only equal 3, 7, or 9. From lines 2, 3, and 4, we also know that I − E = E; or 6 − E = E. Since there was no borrowing from that column into the tens column (G − G = 0), then 6 − E = E can only have one solution: E = 3.

```
1                     _ _ 1
2  _ 1 _ 5 ) _ 6 _ 6 _ 0
3            _ _ _ 0
4            1 _ 0 6 _
5            1 _ 5 5 5
6                5 0 _ 0
7                _ 1 _ 5
8                    _ _ 5
```

```
1                   T E N
2  O N L Y ) D I G I T S
3            L E G S
4            N E S I T
5            N T Y Y Y
6                Y S G S
7                O N L Y
8                  L L Y
```

We can now fill in some more blanks: Multiplying E × ONLY = NTYYY is the same as 3 × ?1?5 = 1?555. A little more deduction is called for. Since the next digit in the answer is a 5, and since it had to have a 1 added to it, the L in ONLY must be some number that, when multiplied by 3, gives us a number ending in 4. Confusing? Let's run through some of the possibilities:

3 × 1 = 3
3 × 2 = 6
3 × 3 = 9
3 × 4 = 12
3 × 5 = 15
3 × 6 = 18
3 × 7 = 21
3 × 8 = 24
3 × 9 = 27

The only solution is 3 × 8 = 24. This means that L must be 8. When we put this answer into our division blank, it checks perfectly. Try it.

Steps

	one	two	three
9			
8			L
7			G
6		I	
5		Y	
4			
3			E
2			T
1	N		
0	S		

```
1                   2 3 1
2  _ 1 8 5 ) _ 6 7 6 _ 0
3             8 3 7 0
4             1 3 0 6 2
5             1 2 5 5 5
6                 5 0 7 0
7                 _ 1 8 5
8                   8 8 5
```

```
1                   T E N
2  O N L Y ) D I G I T S
3            L E G S
4            N E S I T
5            N T Y Y Y
6                Y S G S
7                O N L Y
8                  L L Y
```

At lines 6, 7, and 8, we notice that G − L = L or G − 8 = 8. But the answer isn't 16 (that would be too easy), because, as you recall, we borrowed from the G when we determined that Y = 5. This means that the original number G is 1 more than 6, or 7.

```
1                   2 3 1
2  4 1 8 5 ) 9 6 7 6 2 0
3            8 3 7 0
4            1 3 0 6 2
5            1 2 5 5 5
6                5 0 7 0
7                4 1 8 5
8                  8 8 5
```

Lines 2, 3, and 4 give us D − L = N, or D − 8 = 1. We know that we're safe from borrowing in this case, because in the hundreds column we have already filled in 6 − 3 = 3, and there was no need to borrow for that. So D − 8 = 1, implying that D = 9. So the last unknown. O, must equal 4.

We might have noticed earlier that O and Y are one digit apart by the borrowing rules of subtraction.

Steps

	one	two	three	four
9				D
8			L	
7			G	
6		I		
5		Y		
4				O
3			E	
2			T	
1	N			
0	S			

That took a lot of explaining, mostly because we had to reproduce our logic in words. But once you get comfortable with the patterns in puzzles like these, you can move along, thinking arithmetically instead of in words. In the next few pages, we offer some unsolved Letter Divisions, each with a grid for putting in your answers, step by step. Paul Tobias, who has created these puzzles for us, has provided some clues in the answer boxes to the left. You will find the complete solutions to these puzzles on page 164, but we encourage you to try figuring out the puzzle for yourself.

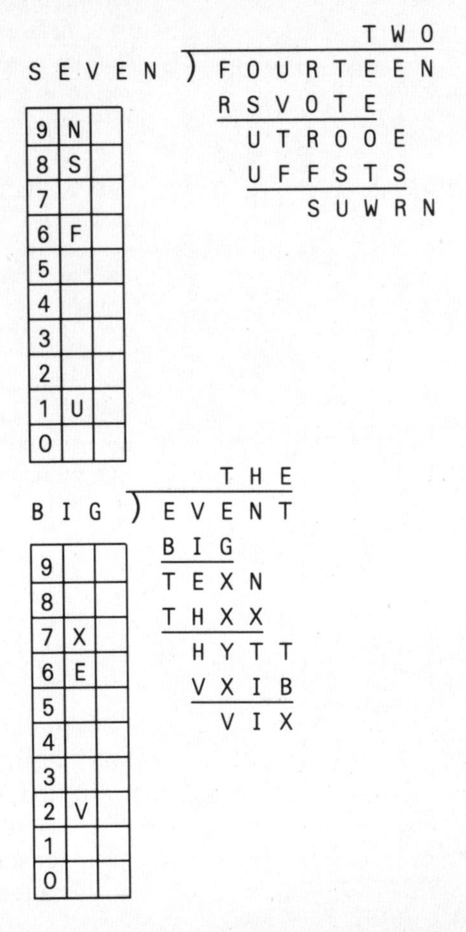

```
                T H E
A-B-C ) O F _ M A T H
        E A   A T
          H   C C T
          C   O A H
              B X E H
              B H F E
                C B T
```

9		
8		
7		
6		
5	E	
4		
3		
2		
1	C	
0	F	

```
                  F O R
W O R K ) S U C C E S S
          S U X W S
              C U S S
              W O R K
              U O C K S
              U X F W X
                O O W S
```

9	C	
8		
7		
6	W	
5		
4	S	
3		
2		
1		
0		

Breaking the Rules

In addition to telling you not to guess, your math teachers may have loaded you down with rules for solving problems. Math-anxiety specialist Susan Auslander, who worked with us at the Wesleyan Math Clinic in the 1970s and today runs a tutoring service in mathematics, statistics, and computer science, takes a very different approach. "A trick that works is a technique," she tells her clients. Though there are hard-and-fast rules (called "procedures") in arithmetic and algebra (e.g., whatever you do to one side of an equation you must also do to the other), there are no "rules" for solving problems, only what works.

Yet we all assume there are rules, even when they aren't given. Take the connect-the-dots problem invented by James Adams and adapted by Paul MacCready, who was the first to build an airplane—the Gossamer Condor—powered by a human pedaling a bicycle. Adams and MacCready want to demonstrate how people block their own creative powers by trying to "follow the rules." Here's an example:[2]

Following is a nine-dot display. The assignment is to connect all nine dots with 4 straight lines, without lifting your pencil from the page.

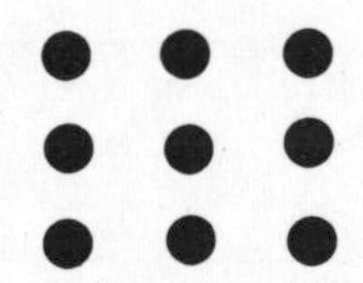

If you follow the "rules," as most people will do, you will need five lines to do it.

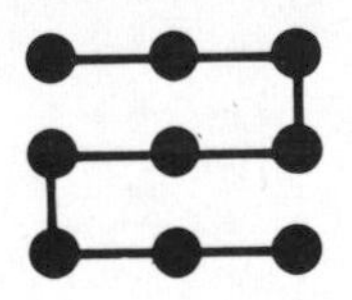

But, by breaking the imagined rule that you have to stay within the boundaries of the dot drawing, you can do it with four.

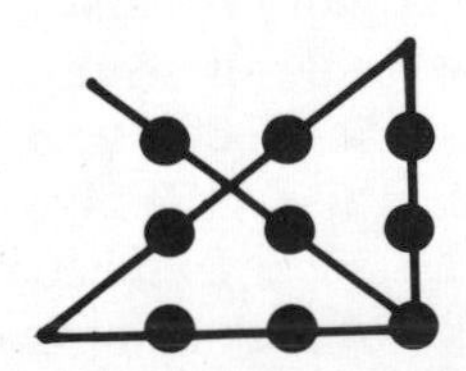

Another assumption people tend to make is that the dots are points without area. But suppose you see the dots as disks having area. This permits you to touch the edges of the dots rather than to go through their centers, as in the drawing below:

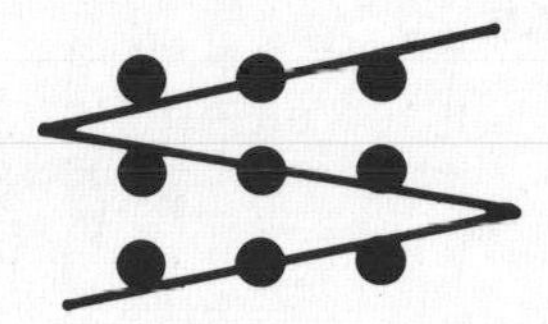

Now challenge another assumption: Why does the line have to be a skinny line? If you can draw the "line" as thick as you want it to be, you can cover the dots with a single "line."

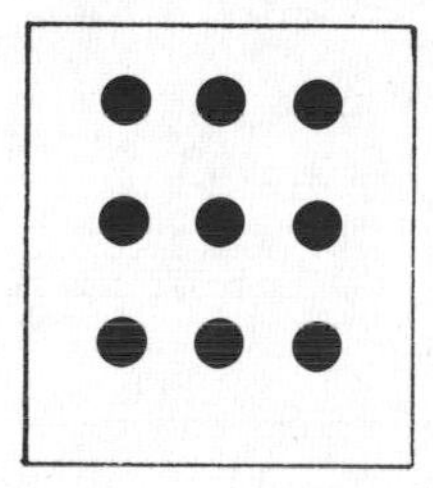

Maybe you can go around the earth a few times until you've connected all the dots.

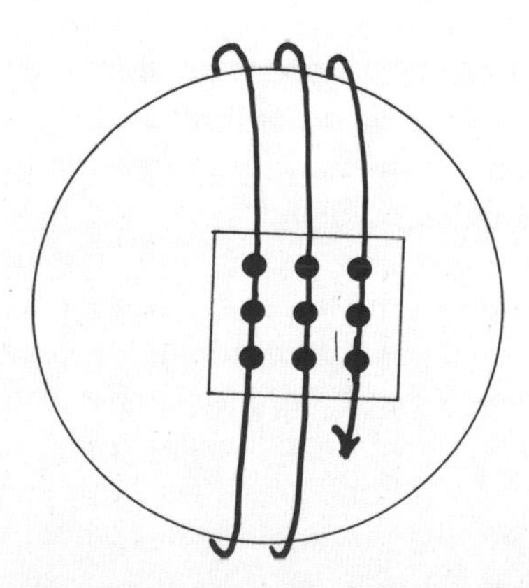

Who says you can't drill holes through the dots, and wrap the dots around your pencil?

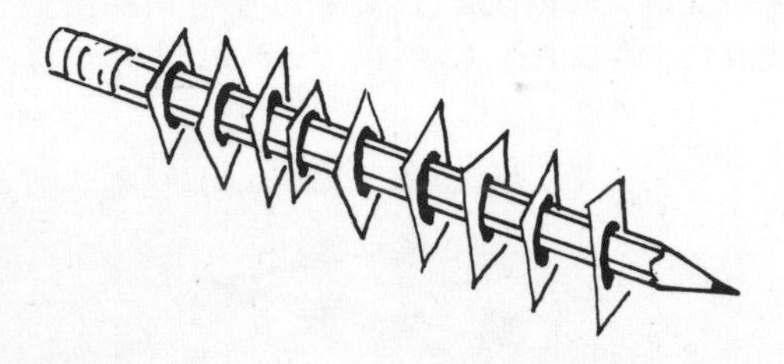

Or crumple up the paper and shove the pencil through it?

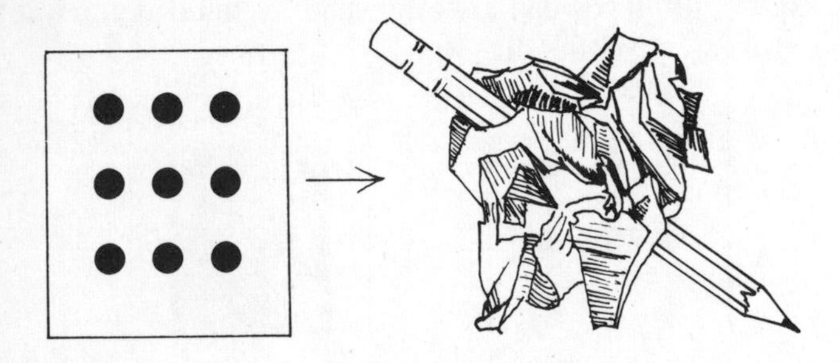

A statistician would say that, when the paper is crumpled, there is a good statistical probability the pencil will hit all nine dots.

Or do some fancy paper folding until you get the dots into one straight line? (See next page)

The point is: Breaking the rules—or "thinking laterally," as the former chairman of the board of IBM used to put it—is not just more creative (and more fun), but often the very best way to think oneself through a problem.

Real problems—the kind we need to solve with creativity—didn't show up much in math class in school. Rather, we had to do "exercises," problems that responded to cookbook methods. But even "exercises" can be approached creatively, especially when extra credit is given for "imagination," instead of being taken away. The Tire Problem, as we have seen, yields to an analysis based on "tire-miles." In Letter Division, you can start

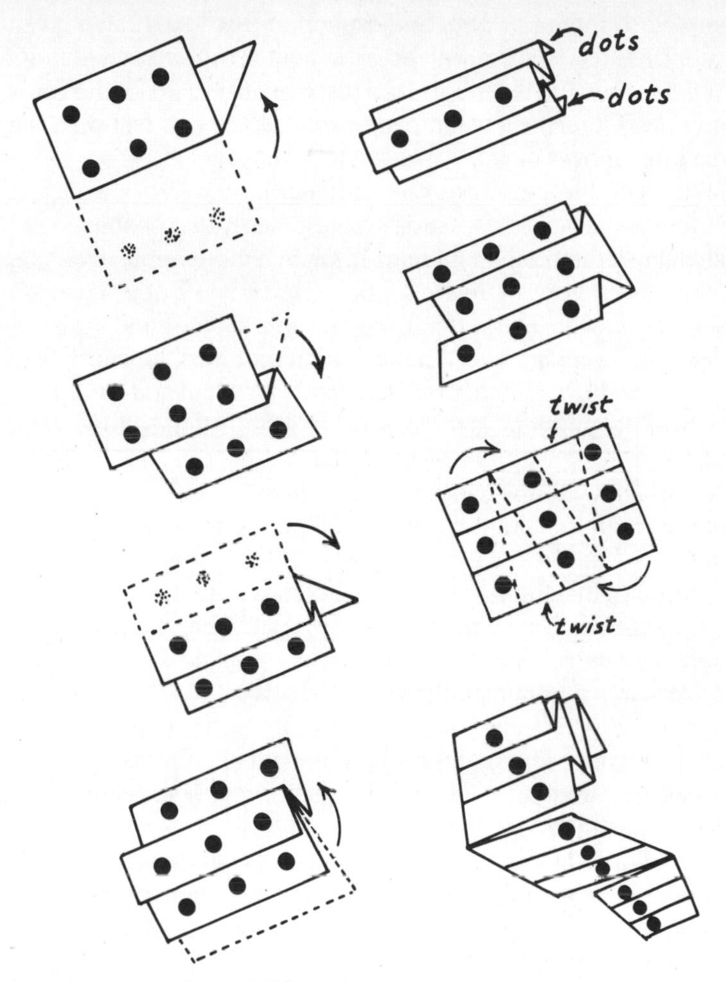

just about anywhere (although starting with zeros and ones is the standard approach).

Conclusion

One of the tricks in solving mathematical problems is not really a trick at all. It involves attending to some of the detail that we are tempted to ignore at first. Some people to whom I have given Letter Divisions to solve will record systematically all the relationships between the letters, and not just dive in and go for the zeros and ones. Others will want to *start somewhere* and will plan the next three moves only if they have to.

Why will one person at one moment pass over a detail and another focus on it? Psychiatrists have a theory that people do not like detail if they cannot manage it. Or, in a more popular version of the same theory, there are supposed to be two kinds of people: sharpeners, who make detail even more visible, and levelers, who try to carve it away. What is important for us, however, is to understand how a mathematician manages detail and gains control over the problem. Is the process of listing all the knowns and unknowns, of giving unknowns letter designations and relating them to knowns, a way of managing detail? "Let t = the time it takes us both to paint the room," one problem solver suggested; "then . . ." and she went on more confident than before. One theory has it that the word "algebra" comes from the Arabic word *algebrista,* which in Spanish also means "bone-setter." The algebraist, with his transformations and manipulations, can "set" a problem into a form that allows him to solve it.

From this perspective, then, panic in the face of apparent ambiguity might be just another way to feel out of control. Yet detail, instead of muddling up the issue, might provide reassurance, if one could attend to it constructively.

Sometimes the problem isn't too much detail, but too little. To know when one has enough information to solve a problem is a problem in itself. Take this one, which a group of mathematicians took about as long to solve as anyone:

Two college roommates meet after many years. One asks, "How old are your three daughters?" Answer: "The product of their ages is thirty-six." Questioner: "But that's not enough information." Answer: "Well, the sum of their ages is the same number as the post-office box we shared at college." Questioner: "That's still not enough information." Answer: "The oldest one looks like me." Questioner: "Oh, now I know their ages."

If you can figure out why that last remark gives the questioner enough information, then you can figure out their ages.*

Partial information is by no means peculiar to mathematics, but, as we have seen, some people have a lower tolerance for "ambiguity" in math or science than in poetry, history, or foreign languages. Is this because they have already learned to distrust math and to doubt their ability to extrapolate from partial information to the knowledge they need? Or can they be taught to cope with partial information in every field?

Some educators are beginning to teach problem solving (heuristics) independent of any particular field or discipline. Others think that certain exercises, like sketching or working with concrete materials, can effectively ready people for problem solving in math. It is by no means certain that math-problem solving can be taught in the same way as facts or interpretation of material. But there is no doubt in my mind that facility in solving word problems can be increased with practice. In the math clinic I have been associated with, people are advised to do word problems every day, like morning exercises, to add to their repertoire, to increase their exposure to all types of problems, and above all to reinforce their self-confidence.

One of the most obvious differences among individuals in attacking word-problem solving is their willingness or unwillingness to keep at it, to pick it up after an interval and try again even if an apparent solution turns out to be false. Some people's attention is disrupted by failure. Others have just the opposite reaction: The joy of mathematics is precisely the challenge of it, the feeling that it is there to come back to, and the knowledge that there is no need to finish it in one sitting. Failure, or lack of immediate success (which is quite another way to think about failure), bothers them not at all.

The question is of course whether, owing either to innate and ingrained characteristics or to role socialization, some people are less likely than others to develop the appropriate tempera-

*See answers for this chapter, page 165.

mental characteristics for solving word problems; or whether the experience of success itself makes people more tolerant of failure.

Maybe there is no "right temperament" at all: Mathematics has some standard moves, many rules and facts and proofs. But as I observe myself and others doing math, I notice that we can make ourselves aware of our own perceptual and strategic preferences, the kinds of models we prefer, the way we approach problems best. Thinking about problems leads inevitably to thinking about thinking. In the Tire Problem, one person likes to get a picture of the problem in terms of the smallest unit, the number of miles one tire has gone during one cycle, or the number of miles one tire will be in the trunk. Another will invent a unit to help him get the whole picture. Or, in the Painting-the-Room Problem, one person will translate it from hours into minutes to avoid dealing with fractions, whereas another prefers hours and even fractions of hours because big numbers make her nervous.

Such individuality could be acknowledged and encouraged, although it plays havoc with lesson planning. In mountaineering, climbers learn to move in terms of both the mountain and themselves. Strong arm pull, good or poor natural balance, preference for chimneys, and tolerance of exposure on the mountain are considered in deciding where and how to move up. Learning to do math, like learning to climb, involves above all learning about oneself.

Answers

Page 141, the upstream-travel problem: The boat takes 3⅓ Hours (3 hours, 20 minutes) to make the 10-mile trip upstream.
Pages 156, 157, Letter Divisions:

```
                        7 2 0                1 3 6
                        T W O                T H E
                  _____________              _________
S E V E N  )  F O U R T E E N      B I G  )  E V E N T
8 3 4 3 9     6 0 1 5 7 3 3 9      4 5 9     6 2 6 8 1
```

```
             7 2 5                       7 1 5
             T H E                       F O R
A-B-C ) O F _ M A T H    W O R K ) S U C C E S S
8 4 1   6 0 _ 9 8 7 2    6 1 5 2   4 3 9 9 8 4 4
```

Page 162, guessing the daughters' ages: A good way to solve this problem is to lay out the possibilities after the second question has been answered in a format like the following. What groups of numbers multiply to 36 and how do they add up?

POSSIBILITIES	PRODUCT	SUM
6 × 6 × 1	36	13
9 × 4 × 1	36	14
18 × 2 × 1	36	21
3 × 2 × 6	36	11
3 × 3 × 4	36	10
12 × 3 × 1	36	16
9 × 2 × 2	36	13

We must assume that both women remembered the post-office box number. If it had been 14 or 21 or 11 or 10 or 16, the problem would have been solved when the answer to the second question was given. Since that was not enough information, the post-office box number must have been 13, which meets both the multiplication and the addition requirements and "is not enough information."

Two possibilities remain that both add up to 13: 6, 6, and 1, and 9, 2, and 2. When the first woman says, "The oldest one looks like me," she is giving it away: The oldest one is not a twin. Hence the combination is not 6, 6, and 1 but 9, 2, and 2.

Notes

This chapter has benefited from the contributions of Nancy Angle (guess-and-check), Susan Auslander (the Tire Problem and problem solving in general), Paul J. Tobias (Letter Division),

Mary Ellen Hunt (Letter Division solution-finding), and Gary Horne (developing for print Paul MacCready's connect-the-dots).

1. D. E. Smith, *History of Mathematics* (New York: Dover Publications, 1958).
2. James L. Adams, *Conceptual Blockbusting* (Stanford, Calif.: Stanford Alumni Associaton, 1974), pp. 16–20.

6 Everyday Math

The *Republic,* Plato's best-known dialogue on politics, does not begin by addressing that subject directly. Rather, it opens with a discussion of what it takes to live a just life. This inquiry, reportedly led by Socrates, Plato's teacher, leads artfully to the construction of an ideal community in which justice is found "writ large." Many have noted that the "Socratic method"—the reliance on questioning and dialogue rather than exposition by argument—works as well as it does because the participants start with what they already know in order to make new discoveries. By this method, Socrates can start almost anywhere, much as if he were throwing a stone out into a river and pursuing the ripples it produces. The result is that the participants are stimulated to think and educate themselves. Even as readers, we may not agree either with the definition of justice or of the ideal society that emerges from the dialogue, but by following the argument we can develop a better understanding of the subject and a better awareness of our own views.

In the same way, the sequence in which elementary mathematics is taught may not be the only way to learn it. Instead of starting at the beginning, wherever that might be, we could start somewhere in the middle, with an interesting question that inevitably will bring us around, if not to the beginning, at least to other interesting mathematical ideas. Adults, whose conceptual equipment is already fairly sophisticated, might best learn elementary mathematics the second time around by diving in somewhere, anywhere at all, and, assisted by an informed interlocutor, proceed in ever-widening concentric circles.

To demonstrate this approach (which is, incidentally, the one I use), I have selected four topics, three in elementary mathematics and one fairly advanced, to treat not as steps in a hierarchy of accumulated knowledge but as points of interest along the way. I assume, as Socrates did, that my readers are experienced, able to think for themselves, and eager to understand. In this chapter, the way is somewhat familiar because most of us have studied these topics before. The material in chapter 7, in contrast, will be new to many.

The topics I have selected interest me. They are: fractions, the many meanings of "minus," averages and averaging, and (in chapter 7) calculus. One probably cannot go very far in mathematics without learning computation skills, but perhaps one needn't begin with them. Because of this belief, I am prepared to dispense with the normal sequence in learning mathematics.

Let's give it a try.

Fractions the Second Time Around

> Arithmetic is usually taught as all scales and no music.
>
> —Persis Herold

Why is it so difficult for adults to remember or figure out how to go from a fraction to a percentage? This topic comes up again and again in workshops for math-anxious people. "How do you get a percentage out of 7⁄16?" "How do you know what to divide into what when you are told that a man pays so much in taxes, at such and such a rate, and you need to deduce his total income from that information?"

I have been wondering about these questions for years and, in the process, teaching myself fractions all over again. My conclusion is that fractions were never easy. It might have been wiser, as the French do, to postpone teaching fractions until they are absolutely necessary in algebra. The French have their pupils do all their "fractional" arithmetic using decimals. But since we rely less on decimals than do the Europeans who long ago adopted the metric system of measurement, we introduce fractions, in my opinion, too early, when most children do not have the capacity to understand them in any depth.

My attempts to make sense of fractions have taught me that specialists themselves disagree about how to present fractions, even how they should be defined. One specialist (Persis Herold) finds it useful to distinguish several different *meanings* of fractions. A mathematician who has become one of my trusted advisors (Sanford Segal) thinks it's better to convey to students that *division* is the central meaning of fractions, although fractions come in various contexts. Whatever the solution may be, the problem is obvious once you interview math-anxious adults.

Most people remember sailing through addition, subtraction, and even multiplication of fractions (although, again, it was more than a little off-putting when multiplication produced *smaller* quantities). They didn't really have to think about what fractions *meant.* When they got to division of fractions, the fact that they didn't know what they were doing finally got in the way. Most

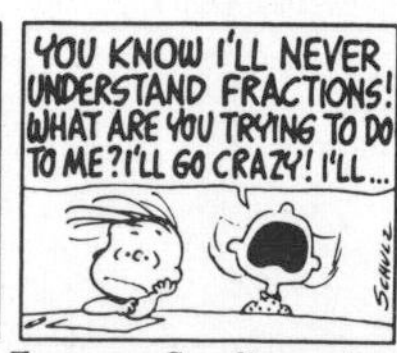

math educators agree that we do not have to *understand* fractions to manipulate them, but I say we have to have a sense that their operations follow some logic if we are going to feel secure about using them.

I would begin with the 4 *contexts* in which numerical fractions appear. Bearing in mind, as Segal says, that the central meaning of fractions is division, ¼ is sometimes presented as if it means one-quarter of one—that is, a *certain part of a whole* (figure 6-1).

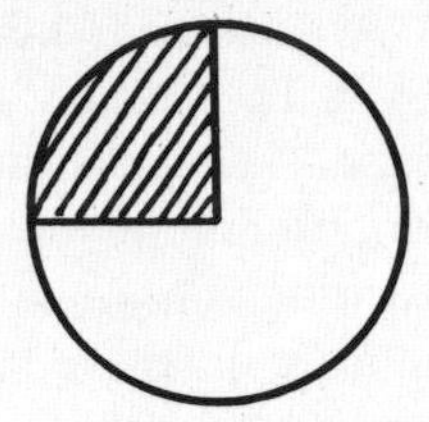

Figure 6-1

But ¼ can also refer to 1 item in a *group* of 4, as in figure 6-2.

Figure 6-2

Third, ¼ can refer to a comparison or a ratio: 1 item *compared* with 4 others (figure 6-3).

Figure 6-3

And, finally, ¼ represents the result of a division; in this case, the fraction means "4 divided into 1" (figure 6-4).

Figure 6-4

In the rate problems in chapter 5, we needed a way to express painting the room in so many hours and we (correctly) used fractions: 1 room in 4 hours was expressed by $\frac{1}{4}$; 1 room in 2 hours was expressed by $\frac{1}{2}$; and the combined rate, the fraction $\frac{3}{4}$, told us that 3 rooms could be painted by 2 people working together in 4 hours (1 room in 1 hour and 20 minutes).

As I remember, the only picture I had in my mind as I was learning fractions the first time was figure 6-1, the pie picture. It would have helped me to be given the idea that all the operations of fractions are rooted in the central meaning of *division*.

Let us examine figure 6-4 a little more closely. If $\frac{1}{4}$ means "1 divided by 4," then going from a fraction to a decimal or a percentage is nothing more (or less) than extending the meaning of the fraction—not a *change* in the fraction at all. I think this is one of the great truths people who have trouble with math never get. The process of "transforming" $\frac{7}{16}$ into a percentage is not a transformation at all, but just another way to state the divisional nature of the fraction.

If we divide 4 into 1 ($\frac{1}{4}$) or 16 into 7 ($\frac{7}{16}$), which is what the fraction *means* to begin with, then we cannot fail to do the percentage transformation correctly. There is no need to ask, "What gets divided by what?" We know it must be the denominator (the lower figure) divided into the numerator (the upper figure) no matter what the size or complexity of the fraction. In the case of $\frac{1}{4}$ (1 divided by 4), the decimal equivalent is .25 (25 hundredths). In the case of $\frac{7}{16}$ (7 divided by 16), the decimal equivalent is approximately .44 (or 44 hundredths). This should work just as well when the fraction is made up of other fractions, such as the fraction

$$\frac{\frac{3}{4}}{\frac{2}{3}}$$

I prefer this notation to the more common

$$\frac{3}{4} \div \frac{2}{3}$$

because it reinforces the divisional nature of the entire operation—things are being divided by elements which in turn are being divided by other elements.

I have asked many people whether and if so when they began to understand all these varied usages of fractions, and to realize that in considering a problem the first step is to decide which aspect of the fraction is going to be useful. Most of my contemporaries cannot remember ever being told this, but I notice that people who succeeded in math managed to figure it out at some point, and that students who had trouble with math did not. Perhaps this long-misunderstood idea—that there are different meanings for fractions—makes otherwise competent people feel insecure with fractions and percentages.

It doesn't help much that fractions are generally introduced to us in terms of pies. As long as we have 1 pie to be divided among 6 people, we can see that there will be 6 portions. This model works well for multiplication of fractions (indeed, it illustrates very well why multiplying fractions may result in smaller amounts), but it does not help us at all in visualizing division of fractions. In division, we want to know not how many *portions* there are but how many *times* 1 portion of the pie will go into the whole.

Division actually means "to distribute," which helps a lot in thinking about division of fractions. Our first sense of division comes in this way. If we have 30 apples and 5 friends and we want to *distribute* the apples among our friends, we divide 30 by 5 and find that each friend gets 6 apples. Fractions are tougher, because we rarely have $^{11}/_{3}$ apples, and more rarely still do we have $^{3}/_{4}$ of a friend. But we can still think about division as a distribution. Suppose we have a board $^{8}/_{3}$ of a foot long and want to divide it (i.e., distribute it) into pieces each $^{2}/_{3}$ of a foot long. How many pieces will we get? (See figure 6-5.) The picture shows that the answer is 4 times. But how do we arrive at that answer mathematically? This one is easy, since we have the same denominator in each case: $^{8}/_{3}$ divided by $^{2}/_{3}$ = 4. But suppose the board were not $^{8}/_{3}$ of a foot in length, but $^{3}/_{4}$ of a foot. Finding the

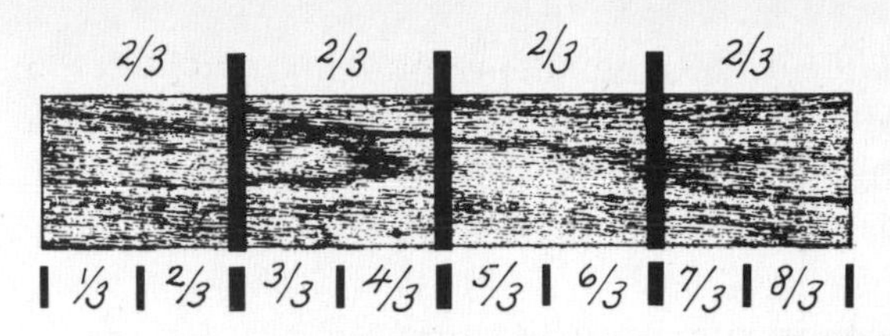

Figure 6-5

lowest common denominator between $3/4$ and $2/3$ yields $9/12$ divided by $8/12$, and the answer is $9/8$ times.

Another way to approach division-of-fractions problems, logically, is to eliminate the denominator altogether, following the rule: Whatever you do to the denominator, you must also do to the numerator.

How will we eliminate the lower fraction? One way is to multiply it by its reciprocal—a fancy word for the fraction that is its opposite. The opposite of $2/3$ is $3/2$. Obviously, when we multiply $2/3$ by $3/2$ we get $6/6$, or 1, and this will nicely take care of the lower fraction. And, according to the rule, we will also have to multiply $3/4$ by $3/2$.

$$\frac{\frac{3}{4}}{\frac{2}{3}} = \frac{\frac{3}{4} \times \frac{3}{2}}{\frac{2}{3} \times \frac{3}{2}} = \frac{\frac{9}{8}}{\frac{6}{6}} = \frac{\frac{9}{8}}{1} \text{ or } \frac{9}{8}$$

Efficiency experts will tell me that I could have achieved the same result by inverting the divisor and then multiplying the upper part of the fraction by that amount. Sure. But I would not have known why I was doing this, and I would have felt insecure about the whole operation. Since what I am *feeling* is as important as what I am *doing*, it is wise, at least for a while, to do what makes sense to me. I think of this intermediate, inefficient method as a kind of scaffolding on which I can rest until the building I am constructing is firmly in place.

My guess is that, after not very many hours of this, people like me will decide not to bother with multiplying both levels by the reciprocal of the denominator (the new rule) but will notice that, since the denominator always disappears when we do that, we might as well shorten the process into two steps:

1. Find the reciprocal of the denominator.
2. Multiply the numerator by that reciprocal.

But when the decision is made to cast off the scaffolding, it will be *our* decision. We will be ready for the short cut and, most important, we will know why we are taking it. In the process, we will also have learned something important about fractions and their reciprocals: that when a fraction is multiplied by its reciprocal, the product is 1. And that insight will be useful to us later on.

Now let's look at what causes trouble in mastering decimals or "decimal fractions," as they are sometimes called. Persis Herold, who has written an extraordinarily useful book for teachers and parents eager to help their children do elementary arithmetic, suggests that confusion with decimals may derive in large part from their layout. Figure 6-6 shows what the various places around the decimal point are supposed to mean.

One space to the right of the decimal point is the "tenths" area; but the "tens" are *two* spaces to the left. *Two* spaces to the right of the decimal point is the "hundredths" realm, but the "hundreds" place on the other side is *three* spaces to the left. Herold has noticed that young people who have difficulty with decimals assume—and didn't we all?—that the decimal point is the midpoint of the number, a sort of fulcrum around which the tens and tenths and hundreds and hundredths are deployed.

But where is the midpoint? Look at figure 6-6 again. It is the area of the ones! No wonder we make mistakes in reading decimals and feel uncomfortable with them.

Though decimals can be converted into fractions, not all fractions are decimals. Fractions can have any denominator: $1/7$, $7/16$, $16/64$, and $16/x$ are all fractions. Decimals have only powers of 10 as denominators, and percentages, which are a subgroup of decimals, have only one denominator, 100. Thinking of fractions,

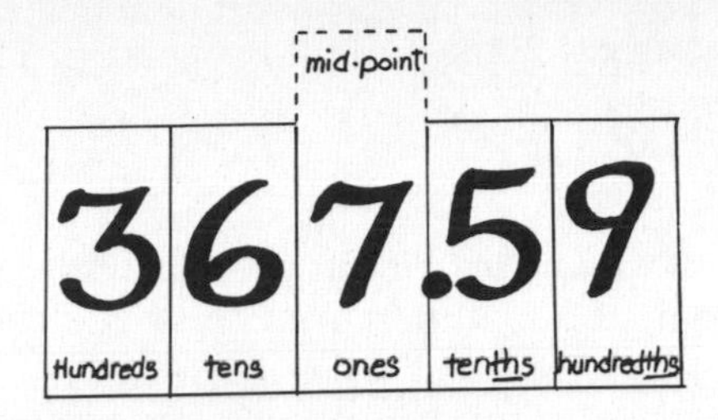

Figure 6-6

decimals, and percentages this way prevents the following kind of error:

> Problem: 6 is what percent of 24?
> Wrong Answer: 6 is 4 percent of 24.

One chooses 4 as an answer because $6 \times 4 = 24$. But the question is really asking what fraction with 100 as a denominator is equal to the fraction $6/24$; since $6/24$ equals $1/4$, and $1/4$ equals $25/100$, the correct answer is 25 percent.

Reviewing fractions this way raises the question: What makes us think something is hard or easy? One reason we may think something is difficult is that we learn it *later* in school; we take sequence very seriously indeed. Since we learn to subtract before we learn to multiply, we assume multiplication is harder. Yet subtraction is harder to do than multiplication. We may think some one thing is harder than something else because of what teachers, parents, and peers *say* about it. Calculus was presented to me as "real mathematics" with such emphasis that I was frightened even to try it. Sometimes a concept is straightforward but its notation is very complex. This holds true in statistics, and even some algebraic notation may cast such a long shadow that an otherwise fairly clear idea becomes murky. A fourth reason, the only one that should matter, is that the concept may actually be difficult to comprehend.

When we relearn elementary mathematics, it is well to sort out for ourselves what is apparently or "ideologically" or notationally difficult from what is conceptually difficult. Take the following kinds of percentage problems, for example:

1. What is 25 percent of 24?
2. 6 is what percent of 24?
3. 6 is 25 percent of what?

It is estimated that 80 percent of all people questioned can solve the first problem; about 40 percent can solve the second; and perhaps only 10 percent can solve the third, especially if the numbers are not as simply related as the ones I have chosen here. One reason the third problem gives more trouble than the first two may be psychological: The problem seems more open-ended. In the first two problems, the largest number is known, and this puts a limit on the range of estimation.

My conclusion from reviewing fractions as an adult is this: Fractions are difficult because they behave quite differently from whole numbers and produce different results when used in familiar operations. Multiplying by a fraction is in fact shorthand for multiplying by the numerator and dividing by the denominator. That's the reason (and children as well as adults need reasons to make sense of things) fractions smaller than one produce generally smaller products; there's a "hidden" division (by the denominator) involved. These smaller products do not mean that the word "multiply" has changed its meaning but that the numbers are a different kind. We, and the children we once were, need to know what *kind* of number we are dealing with in order to know, for sure, which rules will apply.

Another reason fractions are difficult is that eyeballing a fraction doesn't give us as much information as eyeballing whole numbers. Though we ought to know without much thought that 8 is bigger than 7, we do not always know simply by looking at fractions that ¾ is bigger than ⅔. One of the advantages of decimal notation is that comparisons are more obvious. In the case of ¾ and ⅔, .75 (the decimal equivalent of ¾) is clearly larger in quantity than .667 (the decimal equivalent of ⅔). For all these reasons, fractions produce confusion, uncertainty, and even anxiety.

The Many Meanings of "Minus"

If fractions bewitch us, minus signs betray us. For, whereas the meaning of "minus" is rooted in subtraction, "minus" has many usages that are not clearly explained to us the first time around.

We have seen that math-anxious adults are confused when words in arithmetic sound like words used differently in other contexts. Words like "multiply" seemed to change meaning when applied to fractions. Had they been more aware of learning style, the math-anxious would have told their teachers that for them the word "multiply" was simply too strongly associated with "increase" to be helpful in describing mathematical operations.

Mathematicians would agree with such a conclusion. They recognize the dangers of using words in math that have more than one meaning. Indeed, the history of mathematical usage has been a groping for terms that the user would *not* associate with any unintended meaning—for symbols that would not evoke false images. The problem for us learners, however, is that we need a reference point for these words without images; and that reference point is likely to be in commonplace language.

Our first encounter with the minus sign comes rather early in elementary arithmetic. The subject is subtraction, and the format is the following:

$$\begin{array}{r} 10 \\ \underline{-4} \\ 6 \end{array}$$

To make this operation come alive for little children, teachers often use potentially misleading phrases like "take away." This encourages the pupil to think of subtraction as removing one quantity from another.

Subtraction is the first and best-known meaning of the minus sign, but subtraction is a lot older than the sign itself. The minus sign only came into common usage in the fifteenth and sixteenth centuries as a way to simplify and standardize mathematical notation. Before then, other signs were used. The capital letter M, for example, stood for "minus" in English and German, but of course it might have made no sense to the Chinese.

The problem is that the minus sign ($-$) does not designate subtraction alone. It is employed in at least two other ways in school mathematics, and although some pedagogues have suggested that we change the signs to reflect these different meanings, the minus sign is still used for all three.

The second use of the minus sign is to designate a negative number. In my day, negative numbers were actually called "minus numbers," to make it all the more confusing; today the preferred term is "negative number." There are several ways to understand the idea of a negative number. One can *approach* the idea via subtraction: $10 - 4 = 6$; $10 - 14 = -4$. This tells us how to get to -4, but not what it means.

Another approach is to visualize a thermometer (figure 6-7).

This helps us see that a negative number is on the other side of zero and, further, that it is a number in a different *direction* from positive numbers. Another model, the East-West diagram, makes this idea of direction even more concrete (figure 6-8).

Another way to grasp the meaning of a negative number is to think of overdrafts in banking. If we cash checks for more money than we have in the bank, we can accumulate a negative account. The overdraft model is particularly useful in that it makes sense of the idea of *adding* negative numbers: -10 *plus* -4 equals -14.

From this perspective, adding and subtracting signed numbers might better not be called "adding" or "subtracting" but be conceived, rather, as a kind of combining operation. We can imagine ourselves moving backward and forward, up and down, or sideways (depending on the visual model) as we combine negative and negative, negative and positive, or positive and positive numbers. It would then be easier to understand that addition does not *necessarily* result in an increase in value; the operation is, rather, an instruction to "go up" or "go east;" similarly, subtraction no longer means "remove" or "take away" but, rather, "go down" or "go west."

Things are considerably harder to imagine when we want to

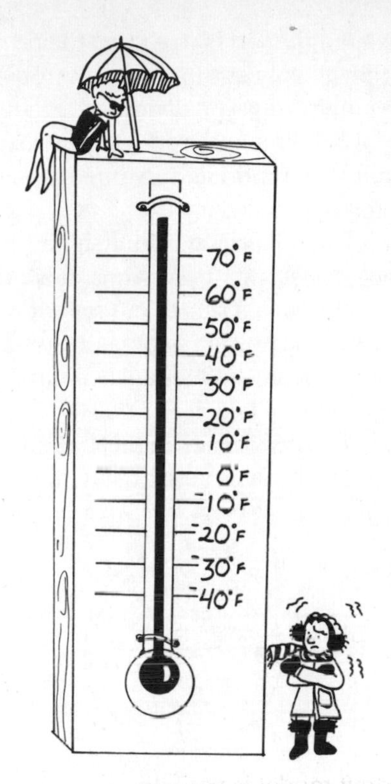

Figure 6-7

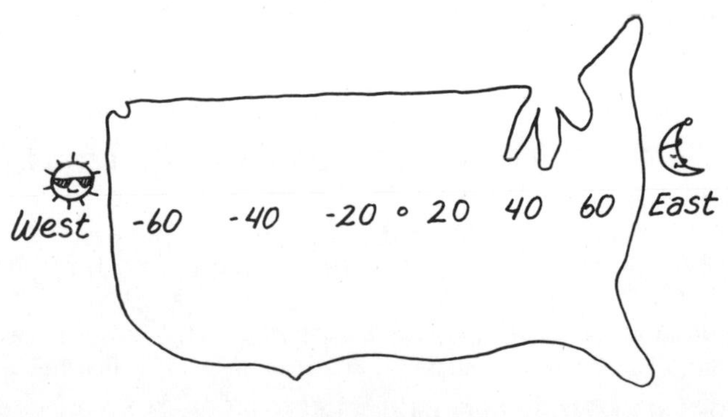

Figure 6-8

multiply negative numbers. It is one thing to add overdrafts, quite another to multiply them. Let's follow this through.

The rule for multiplying (or dividing) signed numbers goes something like this: "Like signs produce positive products; unlike signs produce negative products." By inference, then, there is no operational difference between $(-5) \times (+2)$ and $(+5) \times (-2)$. Yet, when I was learning about signed numbers, since I need to visualize, these two operations seemed very different indeed. Multiplication was a shorthand for addition. I could well understand taking (-5) two times or taking (-2) five times. But what did it mean to multiply *by* a negative number? What do we *do* to something when we take it -2 *times?*

I have found an explanation in an illustration from a history-of-mathematics book that seems very quaint indeed (figure 6-9).*

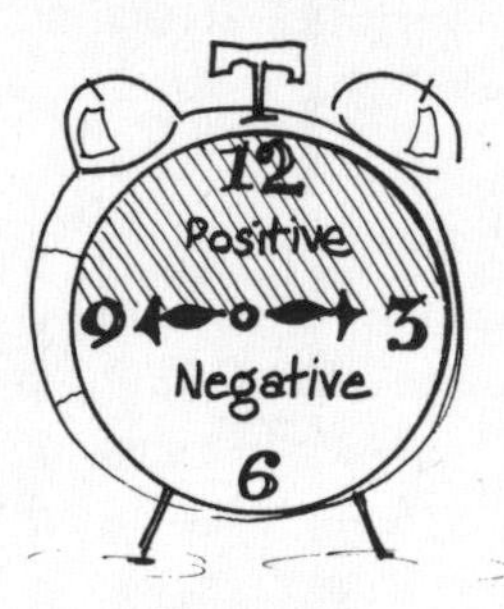

Figure 6-9

The top half (9 o'clock to 3 o'clock) represents the positive realm, where all numbers have plus signs; and the bottom half (3 o'clock to 9 o'clock) represents the negative realm, where all numbers have negative signs. We are to imagine all mathematical calculations beginning at 3 o'clock (another convention), but the direction of turns around the clock is different for like and unlike signs.

If you cannot deal with that model, try the picture my father uses, reproduced in Figure 6-10. But even this, helpful as it is in

*Lancelot Hogben, in *Mathematics in the Making,* describes this model. Pinhead Studios and I have created this particular illustration from that description.

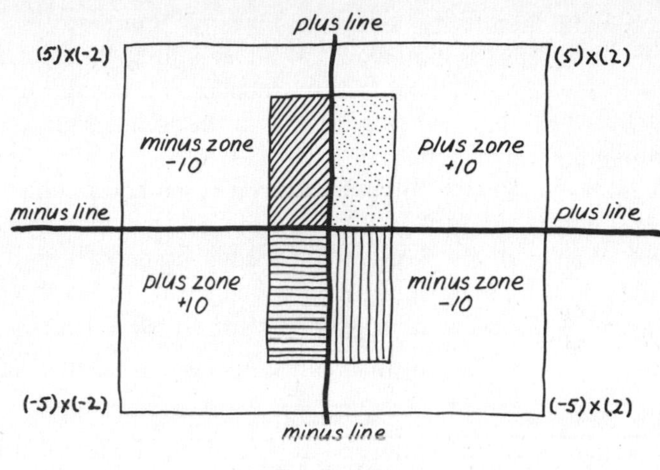

Figure 6-10

reminding us what to do, does not tell us *why* a negative multiplier acts the way it does.

If you are confused, fret not. My only purpose in describing the clock model is to show how very forced the "explanation" is. This is because there may be no commonsense explanation for the behavior of negative numbers in multiplication; the only explanation is that, for arithmetic to remain consistent as a logical system, certain operations must produce certain results.

Once we understand that the system is designed for our convenience, we may not be as disturbed by it as we were before. In arithmetic, the multiplication of positive and negative numbers must produce negative numbers even if the meaning remains obscure. Again, we have to suspend disbelief, because we are not in a realm where the measure of truth is reasonableness. In math, the test of all reasonableness is consistency!

Still another use of the minus sign is in the expression $-x = 6$. Negative x may look like just a negative number, but it isn't necessarily, because x itself may have a negative value. When $x = -6$, then $-x = 6$. The minus sign here is telling us to "change signs." Thus, in one instance, this minus means "choose the opposite." One person has suggested that for $-x$ and similar expressions, the minus sign be replaced by an "°" and $-x$ written instead as $°x$. This notation would avoid an entirely different use of the minus sign, but (with a few exceptions) it has not taken hold, so we are left with a third use of the minus sign, quite different from the other two.

My problem with x^{-2} might have been eased had I known of this meaning of "minus." In this case, the negative indicates an inverse—that is, that x^{-2} has an "opposite," namely x^2, and the two multiplied together equal 1. In the same way, x^2 has another "opposite" which when multiplied times x^2 equals 1. That opposite is $\frac{1}{x^2}$. Thus, it is less surprising that the two "opposites" of x^2 will be the same.*

A further cause for confusion is that in elementary school, when we encounter these apparent inconsistencies, we are not told that a symbol may have more than one meaning. The equal-sign symbol itself has two meanings: that the quantity on one side equals the quantity on the other (as in $6x = 64$), and that the name of the thing on one side is the same as the name of the thing on the other one (as in $x^{-2} = \frac{1}{x^2}$). Had someone shown me this distinction between identity and equal quantity years ago, I might have had less trouble with mathematics. Without knowing that it existed, I needed the three-dash equal sign, "≡," which means "means the same as" or "by definition." Having this sign would have made my life easier.

It would be misleading to leave this discussion without some recognition of the enormous benefits of mathematical notation, especially modern algebra. Figure 6-11 shows many ways of writing simple equations just during the two-hundred-year period of the Renaissance.

Note the number of languages used as well as the variety of symbols. One man, François Viète, decided to make the vowels stand for the unknowns and the consonants for the knowns. Descartes chose the first letters of the alphabet for the knowns and

*The following is by no means a "proof." But it is surely a reassurance: x^{-2} times $x^2 = x^0$ or 1. (Signs are added when we multiply) $\frac{1}{x^2}$ times $x^2 = \frac{x^2}{x^2} = 1$.

Development of Algebraic Notation

date	mathematician	symbolism used	modern form
1494	Pacioli	"Trouame .1.n°. che giõto al suo q̄drat° facia .12."	$x+x^2=12$
1559	Buteo	1◇P6ρP9[1◇P3ρP24	$x^2+6x+9=x^2+3x+24$
1577	Gosselin	12LM1QP48 aequalia 144M24LP2Q	$12x-x^2+48=144-24x+2x^2$

The earth is flat!

Figure 6-11

the last letters for the unknowns, a convention still in use today. Modern mathematicians use all the Greek letters and many other symbols, too. Although it does make learning more difficult, the abbreviated notation of modern algebra makes all mathematical operations easier to do. We can see from figure 6-11 how difficult it would be to manipulate equations if the notations were still so cumbersome.

The adoption of generalized symbols was especially important because it freed thinkers to examine and manipulate the operations and relationships they created, whether or not these were realistic or possible. As long as the ancients did not use generalized formulas, they were limited in what they could consider.

Imagine for a moment a time when negative numbers (-4), irrational numbers ($\sqrt{7}$), improper fractions ($^{11}/_{9}$), or imaginary numbers ($\sqrt{-2}$) had not yet been conceived.[1] Then the following equations could *not* have been solved:

$x + 6 = 4$ (because x would equal -2)
$2x = 5$ (because x would equal $5/2$)
$x^2 = 7$ (because x would equal the square root of 7, $\sqrt{7}$)

$x^2 = -4$ (because x would equal the square root of -4, $\sqrt{-4}$)

Expressing these same equations in general terms, however, by substituting letters for numbers, allows us at least to consider solutions:

$$x + b = a \quad x = a - b$$
$$bx = a \quad x = a/b$$
$$x^n = a \quad x = \sqrt[n]{a}$$

The advantage of this generalized notation is that within it one can work even on "illegitimate," "meaningless," or even "nonexistent" solutions. This frees the thinker from irrelevancies, enabling us to consider possible relationships before we have any experience with them. In this way, notation is a language, for we have the words "red" or "father" or "democracy" in our minds before we have any particular experience with them. Generalized notation permits the kind of supposing that led Newton to wonder whether things on earth were not naturally at rest, but naturally in motion, and, if they are in motion, why they should ever stop. Astronomers could predict the existence of the planet Pluto before it was ever observed through a telescope, and physicists could predict the existence of certain nuclear particles long before they were isolated, because they bore a mathematical relationship to particles already known to exist.

We are left in a quandary. Do we name impossible things with impossible words, thus avoiding any confusion between the language of math and the language of everyday life? Or do we choose familiar language that is loaded with content? The minus sign is an interesting case in point.

Averages and Averaging: One Way to Think About Statistics

Nothing is as useful or as intimidating as statistics. The "fact" that "three out of four people" do or do not do; that "40 percent of the labor force" is or is not; that "the typical American male" earns or wants or believes, is very familiar rhetoric today. Our president's staff includes his own polltaker, to keep him informed of what the public is thinking. Commercial polling has become a large and prosperous industry.

We are intimidated by statistics, not only because they seem to denigrate our intuition but because those of us who are uncomfortable with figures believe that statistics is a science of numbers and relationships in which human judgment is unimportant. In fact, statistics is no more and no less than a substantial improvement on simple averaging. However complex and computerized the techniques may be, reliable and useful statistics must depend heavily on human judgment.

Take averages and averaging as a case in point. We can add up some measurable instances—say, the heights of a group of children, the sizes of a group of objects, the incomes of a group of families, or the amount of June rainfall during the past four years. Then, dividing that sum by the number of children, things, families, or Junes, we can find the "typical" height, size, income, or June rainfall. Averages give us a sense of things, a feel for the central tendency or the trend.

But averages have severe limits, too. The hardest thing to do in dealing with averages is to get out from under the generalizations they suggest. Once we have a statistic, we must still decide whether the information it contains will help us. This calls for judgment based on our knowledge of the situation and the purpose for which we are gathering the data. It may require us to calculate other statistics before we can decide whether to use the information we have; or it may require us to use another technique to test the reliability of our "facts."

Having added all the heights of children in the second grade at a particular school, can we predict the size of next year's second-graders accurately enough to order new chairs? How useful would it be to know, for example, the average income of people living in a metropolitan area? If we add their incomes and divide by the number of families, we will get an arithmetically correct answer—say, $12,000. But that may not be quite so useful if we know in advance that there are a large number of rich people who earn considerably more than $12,000, and an even larger number of poor people who earn less. In fact, we may not want to know "average income" at all, but how many people earn more or less than some amount. Depending upon the use of our statistic, it might be more informative to know that half the population earn about $7,000, a quarter earn $10,000, and the richest quarter earn around $25,000.

Different uses call for different techniques. If our goal in this study of incomes is to understand how a particular family compares with all other families in the population, we will want still another kind of average, a picture of how far or near this family is from the midpoint, if all families were ranked in order of income.

In addition, we need techniques to deal with unreliable data. We must make sure that our heights, rainfall, number of children born, or reported incomes are typical or even reasonable. Averaging is based on the assumption that, if enough incidents are gathered, some typical pattern will emerge. If we use unusually high or low examples because the sample we chose to measure had some oddballs in it, our picture will be distorted. Here, too, human judgment is at work. We approach the data with a certain expectation of what is typical. We have already calculated the average, nonnumerically, in our minds. If the average comes out high or low, we examine the data to see if atypical examples were counted. We cannot automatically throw these out. After all, they are data. But if we can prove to ourselves that they are there because of an unusual circumstance—and here again our judgment comes into play—then we can dispose of them.

One such example occurred after the mandatory deposit on bottles and cans was made a law in the state of Oregon. The state government decided to check thirty segments of highway on a regular basis to see how much, if any, less litter was being thrown out of car windows. The nonnumerical expectation was of course that people would save their bottles and cans for deposit redemption. But would highway litter show a decline? The segments were checked, and the information was put on computer and averaged. The data indicated that twenty-five of the thirty segments showed between 60- and 90-percent reduction in the amount of litter; two showed a reduction of between 20 and 40 percent; one showed an increase of 1 percent; and two had increases of 250 percent.

The last two segments were so out of line that an average of litter per segment might have misled the policy-makers. Possibly something had changed in these locations over the year.* But, if one is to generalize from the sample of segments, the segments have to be representative. Statisticians have a way to deal with this kind of problem. Using formulas lay people do not have to employ, they measure the data's confidence limits, which indicate how much confidence one ought to have in one's data.

Sometimes one wants to know not just the average but the amount of variation around the average as well. It was not very meaningful to know that the average income for a family in a metropolitan area was $12,000. We might want to know the range, or what is technically called the "interval," too, from $3,000, the lowest income, to $35,000, the highest. We might even want to know something more: how frequently various incomes occur. This kind of work is at the heart of social science: finding the "frequency distribution" and the "amount of variation from the mean (or average)." This is so important because we need to know whether our limited picture is likely to predict the next set of segments, or the larger population, or the next day's weather.

What we really want to know is: How tall are the next three second-grade classes likely to be? And will we need to have higher-backed chairs? If the last thirty second-graders have averaged four feet but in an interval between three feet six inches and four feet six inches, the average alone does not give us the whole story. But if we include information about the range and frequency of the deviation from the average, we may be able to order our chairs with confidence.

So far, we have been talking about static averages: snapshots of the typical American family or a typical June rainfall pattern or the size of a typical second-grader. But our questions may require

*In this case, it turned out that two overzealous college students, hired to check segments, were counting orange peels and other degradables as litter. The lesson here is that the criteria must be clearly spelled out.

us to predict trends that change over time, moving averages, things that alter course as other elements affect them. To find the amount of energy required to heat or cool a typical home in Wendover, Utah, for example, the average temperature for the year will be misleading. It is about 51.9 degrees, which would require little heat and no air conditioning. Even knowing that the interval is between 28 and 75 degrees does not help much. The picture of energy requirements has to be a moving snapshot, with averages by the month or even by the week, lined up one after another.

But how lined up? In figure 6-12, note that Eureka, California, with an average temperature of 52.2 degrees, ranges only from about 45 degrees to 56 degrees over the year.

Sometimes moving pictures are expressed in bar graphs, sometimes in line graphs, sometimes as curves. The distinction between bar graphs and line graphs is that bar graphs usually represent discontinuous change and line graphs usually represent continuous change. The static picture in a continuous sequence is written as a point, with all the information about it contained in the location of that point on a graph. The moving picture is a series of points, connected with a straight or, if need be, a curved line. If no line can be drawn because the points are so scattered, then another technique is used to find out what line most approximately links these points.

Say we wanted to get a picture of how the American people are

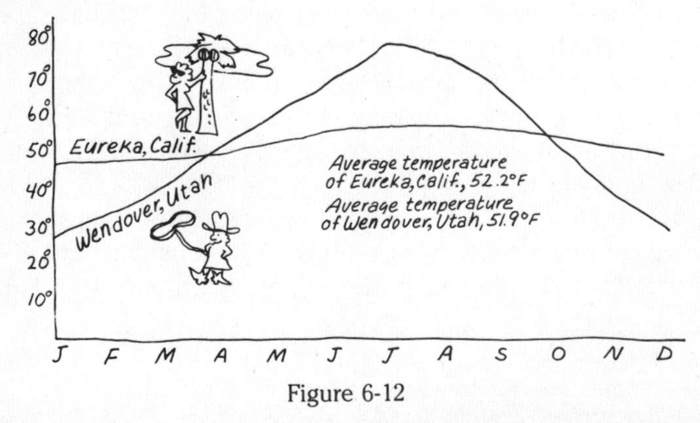

Figure 6-12

doing financially, compared with how they have done in the past. To do so, we would need to look at a number of different moving pictures.

Figure 6-13, representing the Consumer Price Index from 1960 to 1988, is an interesting place to start. It is quite obvious that the curve of prices goes up rather steadily, particularly after the oil crisis of 1973. Note that, for the sake of having some criterion, the year 1982 is selected as "equal to" 100—not 100 anythings in particular, just 100—so that we can call a low point 30 and a high point 110. What the reader of this graph wants is comparisons, not exact numbers, so the arbitrary designation of 1982's prices as "100" is perfectly all right.

The Consumer Price Index would, by itself, suggest that things are not going well at all, for the price of goods and services has increased 300 percent since 1960. If hourly wages and salaries had stayed the same, it would take three times as much work to buy goods in 1988 as it did in 1960. To get a total picture, we also need to know how incomes fared in the same period. Figure 6-14 gives two approximations. The solid line gives a moving picture of the per-capita gross national product (GNP). This is not exactly a measure of income but, rather, of the total amount of goods and services produced by the country, divided by the entire population. The dotted line, plotted for the same years, shows a parallel curve for average salaries in one profession, public-school teaching. It is interesting to note how, from 1940 on, the per-

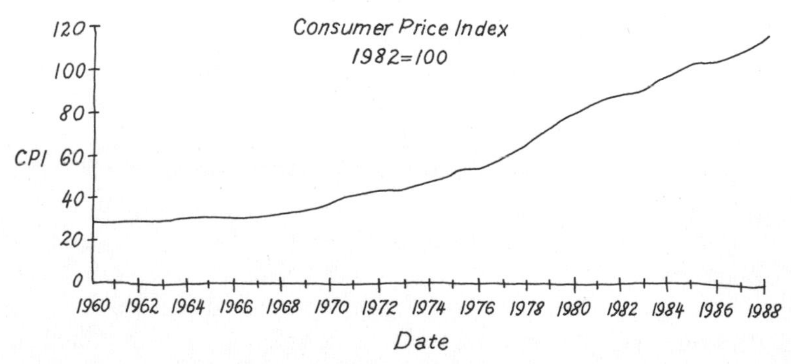

Figure 6-13

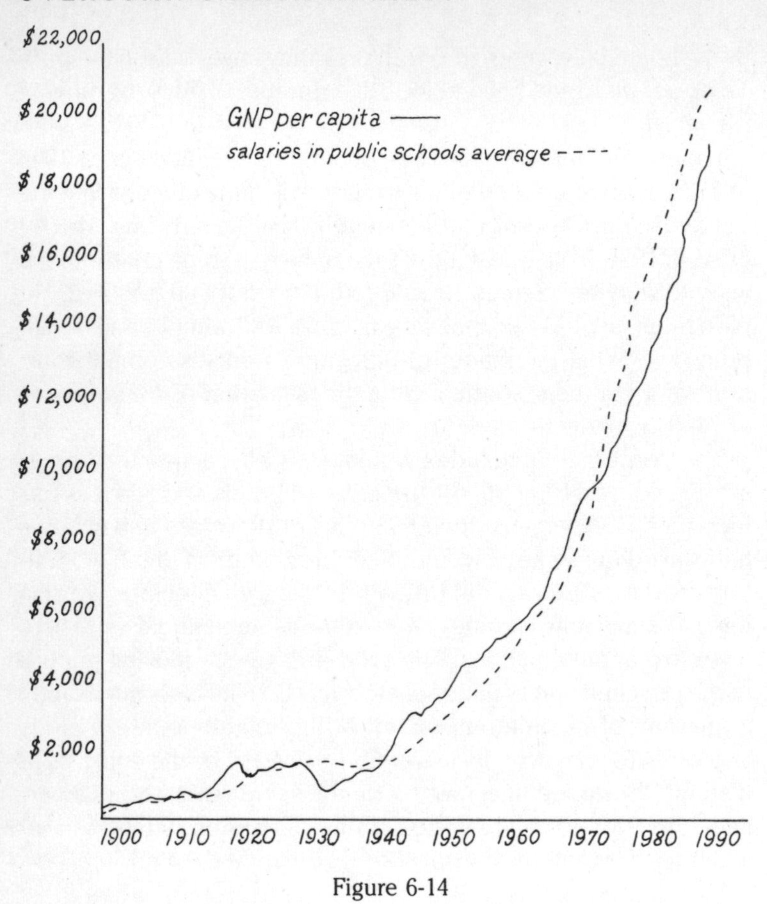

Figure 6-14

capita gross national product increased at a very steep rate, and how salaries, at least in one sector, increased at about the same rate.

Again, amounts are interesting, but even more interesting are patterns of movement, and these can only be compared by drawing line graphs that measure not just quantities but change in quantities.

Our conclusion from these three graphs might be that the American people, on the average, are keeping up with inflation. But what about the others, people on fixed incomes and the unemployed? We might want to look at the unemployment rate over the same period to get a feel for how well some of these other people are doing. Figure 6-15 shows the percentage of the labor force that has been unemployed over the past eighty years. Note the moderate range of variation except for the Depression years. Amazingly, despite the changes in productivity and the increase in the absolute numbers of the population and in the numbers of people, especially women, working outside the home, the unemployment rate has remained almost constant. Critics of the statistical methods used to generate these line graphs of unemployment complain that the way we define the unemployed—as "people actively seeking work" at any one time—causes us to overlook large groups of people: those not able to work because of ill health or inadequate training (the so-called chronically unemployed); people who have given up looking for work in discouragement; and those who for some reason are ineligible for unemployment insurance, like farm workers and other seasonally employed people, or domestics. The absolute number of people unemployed is probably far larger than the unemployment rate, as shown in figure 6-15.

Another interesting statistic is the average age of women bearing their first child in the United States over the past fifty years. What does such a statistic tell us about the reasons for increases and decreases in that age? As is often the case, the statisticians have prepared several interesting numbers, but none that answer our question directly.

Figure 6-16 shows birthrates in terms of age of mother for one

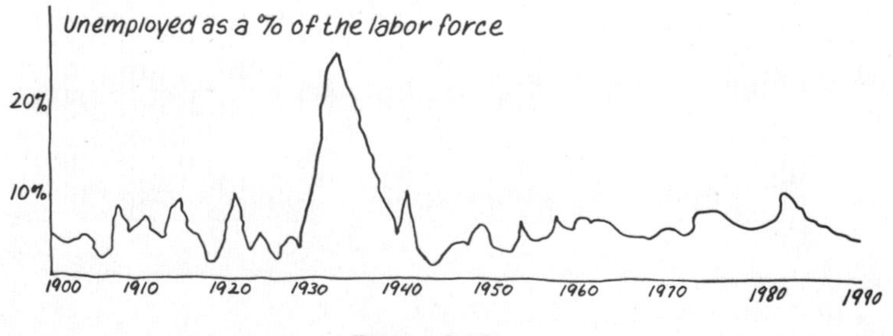

Figure 6-15

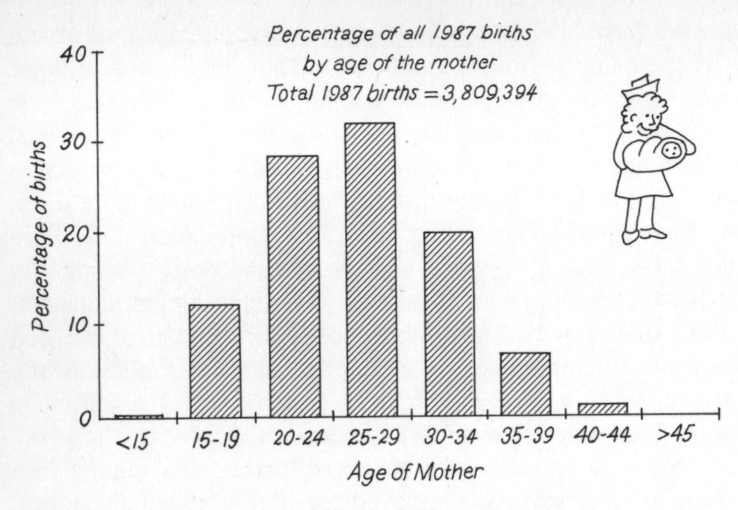

Figure 6-16

year, 1987. What is interesting about this display is what it tells us and what it does not. It tell us the percentage of all 1987 births of mothers in a certain age range. It does not tell us whether these are firstborn or later children—only the age pattern of their female parents as children come into the world.

Figure 6-17 offers us additional information, the number of

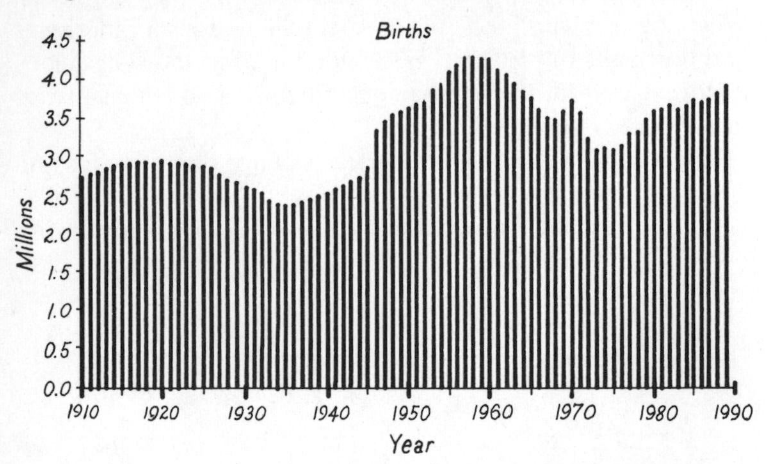

Figure 6-17

births in millions over the time period 1910 to 1986. It would not be difficult to go from this bar graph, which represents quantities, to a curve representing change over time. Even with the changing total population, which means there are more and more women of childbearing age over the period, we can pinpoint the declines very accurately from a graph like this one: for one thing, the 1930s were low in births, the 1940s and 1950s especially high. We call this the "baby boom" and the children born in those years "baby boomers." After 1960, there was a precipitous decline. Why might this be? A social scientist might look for changes in attitudes and values, but the answer is actually in the invention and wide distribution of the first very reliable birth-control pill. Why, then, the modest upsurge in the 1980s? Prosperity? Later marriages? Were couples who had not had babies in the 1970s becoming parents a decade later? Only detailed survey questionnaires could answer all these questions, but rates of marriage and divorce might give us some clues. (See figures 6-18 and 6-19.)In Figures 6-18 and 6-19, rates are given per 1,000 people. For example, of 1,000 people in 1910, 10 were likely to get married (not *be* married) in that year, whereas, during the height of World War II, 12 were likely to get married. Also, the rate of marriage does

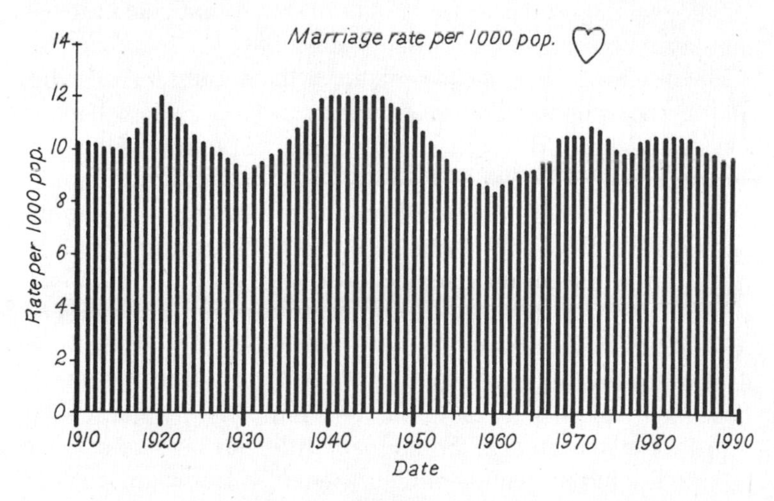

Figure 6-18

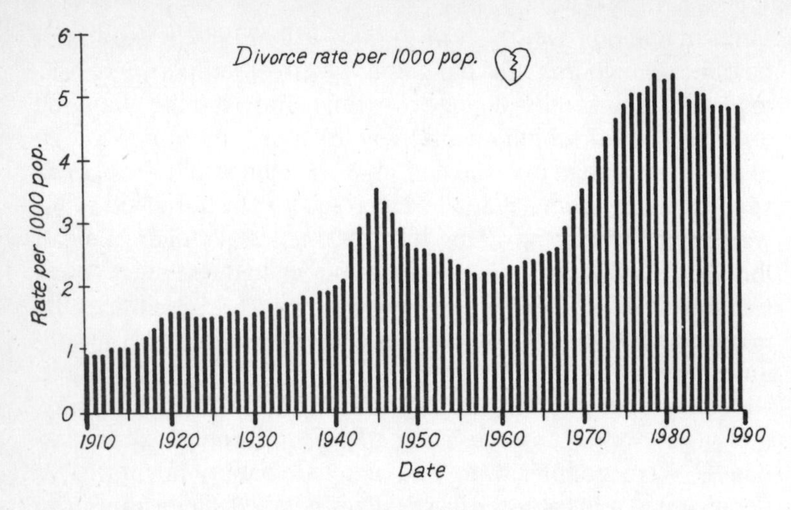

Figure 6-19

seem linked with the number of births, because beginning in 1958 or 1959 the number of people getting married (per 1,000 population) was lower than at any other time in this century. With the number of marriages returning to a previous level in the 1980s, we might (but we also might not) have expected to see an increase in the number of births (and we have).

Divorce rates are a stickier matter. What should the base be? Should we compare number of divorces per 1,000 population per year, as in figure 6-19, or should we compare number of divorces with number of weddings? Or with a base made up of the number of intact marriages at any one time? These bases are important, because the divorce rate compared with the rate of new marriages might seem high just because the rate of new marriages went down. In 1960, there were about 2 divorces for every 8 weddings, or 25 percent. This is a very misleading statistic, because the absolute number of divorces per 1,000 population was down in 1960. In 1946, on the other hand, when wartime weddings were still inflating the marriage rate, the divorce rate was the highest in this century until just recently. But when we compare divorces with marriages that year, the divorce rate doesn't look so bad: about 4 divorces, compared with 14 weddings, per 1,000 population.

The point is that, to get a sense of the quality or the duration of marriage in America, we cannot rely entirely on divorce or wedding rates per 1,000 population. People on the verge of getting married today might be put off by the fact that there are 4.5 divorces for every 9 weddings at the present time, yielding what looks on the surface like a 50-percent marriage-survival rate. But that is not at all what these figures mean. The people getting married this year are not the people getting divorced—a classic misuse of statistics.

Marriage and divorce rates, like the GNP, are very much the product of human choice and human endeavor, and hence rather unpredictable. When we take a statistic from the biological realm, such as median height by age (figure 6-20), we can see how a moving curve rises and plateaus. The graph in figure 6-20 is very interesting in terms of our prejudices about male/female differences. Until age 11, the end of middle childhood, females are either just as tall as males of the same age or even, on the average, a little taller. From age 11 on, girls appear to stop grow-

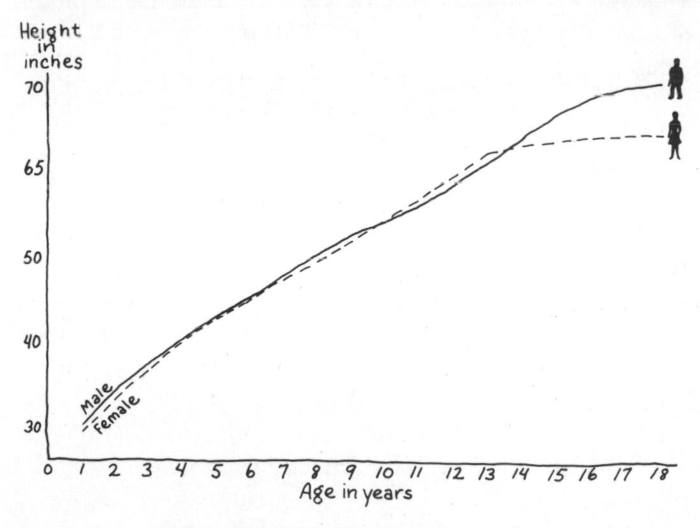

Figure 6-20 Median height by age

ing, at least at the same rate as before, while males continue along generally the same slope. They, too, stop growing, but not until the ages of 16, 17, and 18. Thus, the height difference is clearly the result of a change at puberty, and not a pattern apparent since birth.

The significance of such a picture is that other variables confirm the similarity of boys and girls during middle childhood. The differences in height and strength only show up at puberty. Hence, earlier differences in behavior may be social in origin. There is no reason why an 8-year-old boy should be more interested in and better at sports than a girl the same age.

Figure 6-21 is a scatter diagram. This one is an attempt to show the clustering of language and science aptitudes in a certain university population. Scatter diagrams (or scattergrams or scatterplots) are useful when there are no shapes that can be linked by lines or curves but only clusters of points. Where the points are most dense, of course, one does get some information.

Once one has a line or a series of bars or a cluster of points on a graph, many obvious questions occur that might not have been obvious from the raw data. What accounts for the steepness of the line, for example? Why does steepness increase or decrease at this point? Why is the line straight? Why is it curved?

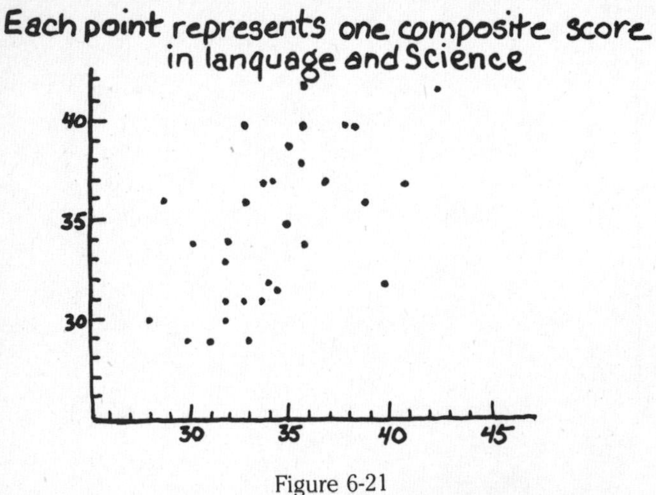

Figure 6-21

Comparing data to look for links and associations among trends or events is an obvious next step. Plotting age at childbearing or numbers of children born to mothers of a certain age against some other development—say, health of the economy—might give one a clue as to *why*. The advantage of the "moving average" is the dramatic image it gives of what happened over time and under changing circumstances. Everything on a graph can be expressed in words, but it would usually take many words to say as much, and the overall impression might not be as distinct.

When we superimpose one moving average on another, we are searching for correlation. The figures we have looked at in this chapter provide some good examples of obvious correlation: the per-capita GNP and the average public-school teachers' salaries; the rate of new marriages and the rate of divorce. To be sure that these are true correlations, we might want further evidence; also, we cannot know whether one trend causes the other or whether they are in tandem because of a third factor, an "intervening variable," such as the influence of World War II on the high marriage and divorce rates in the 1940s. All too often, social scientists who find connections between variables jump to conclusions about the inevitability of their correlation and/or the causative power of one over the other.

"Why" questions can rarely be answered by statistics alone. Why are there more cold days in January than in March? Why has the average age at childbearing for women first gone down and then up? And why has inflation, not a major problem (except during wartime) before 1960, become so severe in recent years? To answer these questions, we need much more than is in the data. We must think up hypotheses, find data describing them, and do more calculations to look for connections between factors. Some forces are quantifiable; many are not. In many cases, careful researchers only conclude that "the data do not disprove my hypothesis." They know that they cannot be sure their hypothesis is true simply by looking at averages and samples of larger populations, however sophisticated their techniques may be.

After one has looked at or drawn many graphs, one can get a feeling for the patterns even before one has put pencil to paper. Age and height of growing children constitute, as we have seen, a picture of upward and forward motion until the plateau is reached at age 11 or 12 for girls and at age 16 or 17 for boys. Being pretty much a straight line, the relationship is called "linear." The relationship between capacity to learn and anxiety, on the other hand, is positive up to a point—one is nervous before a test but should not be paralyzed by it—and then negative. Since this line changes direction, it is sometimes called "curvilinear." If the relationships being plotted have 2 high points or 2 low ones, like the use of electricity in a large metropolitan area, which will have a small peak in the morning and a larger one between 5:00 P.M. and 8:00 P.M., then the curve will be described as "bimodal."

Professional people who look at graphs in their work use a visually oriented vocabulary for discussing issues. The shapes of the curves become themselves a shorthand way of communicating information. "Linear," "continuous," "discontinuous," "curvilinear," "the slope of the curve," and other, similar phrases have become a part of common business parlance. There are no numbers involved in these phrases. They refer, rather, to relationships that have been abstracted from numerical data. And one does not have to be mathematical in order to use them.

At the outset, the newcomer to statistics should decide whether he or she wants or needs a "reading knowledge" of statistics, the ability to get information out of visual or summary statements about data and relationships; or whether the goal is a "performing knowledge," which I take to mean the ability to collect data, select a sample, display information on graphs, and test for reliability or correlations. Few of us will need a "theoretical knowledge" of statistics, an understanding of the theories of probability on which the manipulations are based.

I have learned much about reading graphs by listening to skilled people decode them. People who really need to get something out of this kind of material spend many minutes, sometimes quarter-hours, carefully perusing a graphic representation. Like me, they often begin wrong. They miss the "of" statement or overlook the fact that the data are given in thousands, or based on the 1967 value of the dollar. Graphs are harder to read than sentences, partly because they are not presented in a linear sequence. One doesn't always start at the left and read to the right, and often the key fact or trend is not even obvious until one has digested the whole picture. Ironically, the process is less like mathematics than like pondering a poem. Poetry, too, is not always understood by reading each line in sequence. A poem should be read many times and, ideally, committed to memory. It helps immensely to discuss it and to read it out loud. A graph, too, in my experience, is better understood when it is shared with someone else. Most important is the commitment we make. My prediction, based on a sample of one (myself), is that there is a "positive correlation" between time spent in graph reading and quality of comprehension.

Appendix: A Caveat on Reading Tables

by Allan Johnson, Sociologist

PROPORTIONS AND PERCENTAGES

The *proportion* of people in a group who have a characteristic is the number of people who have it divided by the total number of people in the group. The *percentage* who have it is simply the proportion multiplied by 100. If we have 10 people in a group—8 whites and 2 blacks—the *proportion* white is 8⁄10 or 0.8; the *percentage* white is (8⁄10) × 100 or 80 percent.

We use proportions and percentages to make comparisons, either between groups or within groups over time. In reading percentages or proportions, remember:

1. They must all add up to 100 percent unless some explanation is given.
2. If more than one characteristic is involved in a table, be sure to note the direction of the percentaging (across the rows or down

the columns). Different directions mean that different bases are being used to compute the percentages, and this obviously affects the meaning of the numbers. "Fifty-eight percent of *which group* has *which characteristic?*" This is the kind of question you want to ask.

3. If we percentage across rows, we can make comparisons between columns; if we percentage down columns, we can make comparisons between rows.
4. Before you read a table, study the title to understand exactly what groups and characteristics are included and how they are arranged in the table. Be sure you understand what the numbers in the body of the table mean.
5. Do not take seriously proportions or percentages that have bases of fewer than 30 or so cases.

MEANS, MEDIANS, PERCENTILES

For characteristics that can be added, subtracted, multiplied, and divided (such as income, number of children, years in school, age, etc.), we can *summarize* a whole distribution. A *mean* is an average obtained by adding all the scores (incomes, ages, or what-have-you) and dividing that sum by the number of scores you added (number in the group, etc.). A *median* is the score of the middle person when we arrange people in order of increasing income, age, etc. If we have 11 people and order them by income, the income of the 6th person will be the median—half of the people will have lower incomes than the median, and half will have higher incomes. A *mode* is the most common score.

A *median* is a better measure of the "typical" case, because extreme scores (e.g., millionaires) do not affect its value. The *mean,* however, is sensitive to extremes and is often a distorted measure of the "typical" case. It is very dangerous to compare means and medians, although you will see this done in articles that talk about data from more than one source.

A *percentile* is like a median, but it is more general. The 60th percentile, for example, is that income, age, etc., that cuts off the lower 60 percent of the cases. The median cuts off the lower half (or 50 percent) of the cases, and is therefore the "50th percentile." The 90th percentile is that score below which lie 90 percent of all the other scores. Thus, if the 90th percentile for income is $15,000, 90 percent of the people make less than $15,000 and 10 percent of the people make $15,000 or more.

Note: If my income is in the 80th percentile and your income is in the 90th, then you make more money than I do. We *cannot* tell from this, however, *how much* lower my income is. We could be quite close in income, or we could be quite distant from each other. Thus, if one school district is in the 40th percentile and another is in the 70th, this is no reason to conclude that one is much better than the other. On the other hand, if one school is in the 89th percentile and another is the 88th, they *could* be quite different.

RATES AND RATIOS

A *ratio* is the number of people with one characteristic divided by the number of people with another characteristic. It is a measure of *relative* size. In one example, the ratio of whites to blacks is 8/2 or 4 to 1, meaning that there are 4 times as many whites as blacks. Alternatively, the ratio of blacks to whites is 2/8 or 1/4, meaning that blacks are only 1/4 as numerous as whites.

Note: Ratios measure *relative* size. They do not tell us about *absolute differences.* For example, the phrase "the number of delinquents has doubled" is misleading. An increase from 1 delinquent to 2 delinquents is a ratio of 2 to 1 (a doubling), but it is not very impressive. On the other hand, an increase from 500,000 delinquents to 1 million *is* impressive. Both represent ratios of 2 to 1, but clearly the ratio itself does not tell about the magnitude of the differences or change.

A *rate* is a measure of change over time (mph, population growth, etc.). Many so-called rates (homocide, marriage, divorce, etc.) aren't rates at all: they're static ratios, not dynamic changes over time.

SAMPLES AND SURVEYS

A *population* is any group of people or things that is defined in such a way that we know who belongs and who does not.

Samples are subgroups of a population.

To evaluate a sample, we need to know at least the items below. If we do not, we should not take the survey results too seriously:

1. Whom does the sample represent?
   What population was sampled?
   Did everyone in the population have a *known* chance of being selected?
   How many people who were selected were not interviewed? (What was the response rate? Anything under 75–80 percent is in trouble.)
2. How was the information gathered?
   When was the survey conducted? Where? By whom?
   In what setting was the information gathered? Who did the interviewing?
   How were questions worded? What "yardstick" was used?
3. How large was the sample?

You can use the table below to evaluate many of the poll results that use proportions or percentages. For each sample size in the left column, the right-column figure completes the following statement: "We can be 99-percent sure that the percentage (or proportion) in the *population* is the *sample result* plus or minus —."

So, if our sample size is 1,000 and the percentage of our *sample* in favor of proposition A is 35 percent, we can be 99-percent sure that the percentage in favor in the *population* is 35 percent plus or minus 4 percent—or somewhere between 31 percent and 39 percent. (If we are talking about proportions, the comparable figures are .35, .31, and .39.)

What follows is valid *only* with a properly drawn sample: We can be 99-percent sure that the percentage in the population is the sample result + or −

SAMPLE SIZE	
50	18 percent
100	13 percent
300	7 percent
500	6 percent
800	5 percent
1,000	4 percent
1,500	3 percent
2,000	3 percent
3,000	2 percent
5,000	2 percent
10,000	1 percent
50,000	½ percent

Notes

This chapter has benefited from the techniques Persis Herold introduced me to, and those from the *Math Teaching Handbook* she uses in her classes. On averages and averaging, I used Frederic E. Fischer's *Fundamentals of Statistical Concepts* (New York: Canfield Press, 1973) and personal help from Joel Schneider, now associated with Public Television's "Square One." Two additional books I enjoyed reading in preparing this chapter are the classics: Tobias Dantzig's *Number: The Language of Science* (New York: Free Press, 1954) and Lancelot Hogben's *Mathematics in the Making* (New York: Doubleday, 1960). Gary Horne updated the graphs and provided expert advice for the new edition.

1. This example is taken from Dantzig, *Number: The Language of Science.*

7 Sunday Math: What Is Calculus, Anyway?

The calculus is one of the grandest edifices constructed by mankind.
—anonymous

For those who have not studied it (and even for some who have), the word "calculus" looms large and fierce. Its powers are said to be wondrous, but hard to explain to the uninitiated. The challenge seems to reside in the nature of the subject, its terms, and its fundamental theorems even more than in its avowed difficulty. But, like all great mysteries, it does divide us into those who can and those who don't even dare.

For years I had been circling calculus like a bee attracted by the honey but afraid to land. I had never needed it directly at work, but, as a student of history, I was intrigued that several thinkers had developed "differential" and "integral" calculus almost at the same time. Their challenge, as the history-of-science books put it, was to find a way of *quantifying* the description of the physical phenomena that began to be studied in the seventeenth century—descriptions for which arithmetic and algebra, which were well developed by then, provided no formulas, let alone solutions.

Occasionally, I would bump up against words and phrases associated with calculus, words like "rate of change," "slopes," and "limits," which seemed to enrich my colleagues' capacity to analyze complicated issues—and not just issues related to math or science. Though I often asked for definitions, something about calculus defied explanation, at least in language I could comprehend. The person I buttonholed would either stutter something about these things being too difficult to explain in words, or rush for pencil and paper to draw some squiggles that left me as befuddled as before.

At times I would explore a little on my own, and bought books with titles like *Calculus Made Easy* or *How to Enjoy Calculus.*[1] I made the mistake, however, of trying to leaf through books like these as I would skim a book on any other new subject—say, African art or the history of Byzantium. Even the friendliest preface would soon devolve into formidable examples and esoteric notation. No author seemed able to tell me what calculus is *about* without having me learn how to do it. Yet I was stubbornly determined not to learn how to "differentiate" or to "integrate" simply on faith. It was my turn to be paralyzed by math anxiety. Nothing more, nothing less.

Finally, a mathematician to whom I had confessed my disability intrigued me with the suggestion that, if I thought hard enough, as the medieval monks had, about how many angels could fit on the head of a pin, I might become aware of some of the fundamental issues calculus was designed to deal with. Angels on the head of a pin, indeed. What could that possibly have to do with slopes, integrals, and limits? Think about it, he said.

And so, with beginning excitement, I went back to look at the problem of the angels and the pinhead and followed it as I would pursue any line of inquiry, reading, thinking hard, asking questions of others. Slowly, the angels helped me evolve a way to appreciate calculus, if not yet a way to master it.

For most people, the way into calculus will be through real-world problems such as the rise in actual prices over a year-long period in which the inflation rate fluctuated all the time. Arithmetic will give you the *average price rise in percent* over the year, but only calculus (and your pocketbook) will tell you what the actual accumulated increase was in dollars. Or suppose you wanted to know the rate at which an airplane increases its speed during takeoff. Your pilot will need to know the plane's change in velocity, moment by moment, in order to determine the engine thrust necessary to get the plane safely off the ground. But you can't do that calculation with algebra, nor can your pilot. Questions like these stimulate the kind of inquiry that leads inevitably to calculus. But for me the monks' paradox, far more than the real world, opened the door.

Angels on the Head of a Pin*

The monks began with some assumptions. Although we may no longer share their assumptions, these imposed interesting constraints on the problem. God, they believed, was omnipotent. (This was the first constraint.) Therefore, if He wished, He could create angels of any size, small enough, if He so desired, to fit on any piece of matter, even a pinhead. But because God created the universe according to immutable laws of nature (and this was the second constraint), even He could not create angels that took up no room at all, or a pinhead that would contract and expand as the angels climbed on and off. No, the angels and the pin had to be of some finite size. And so the meaning of the problem posed by the Scholastic was this: Can an infinite number of finite things fit onto a finite space?

We begin, as the Scholastics must have done, with pure logic and sheer, uninhibited speculation. Let us imagine that God started with 1 angel, an angel that alone would fill up all the available space on the head of a pin. God could just as well

*I am indebted to Joseph Warren, then an associate of Mind Over Math in New York City, for this appreciation of the monks' argument. This discussion is, of course my own.

create another, different-sized angel, ½ the size of the first, permitting 2 angels to fit on the pinhead. Another reduction and there is room for 4. More reductions allow 64 angels to climb on. Several more strokes of God's hand and 1,024 still tinier angels are aboard.

Or, instead of creating a whole lot of angels ever smaller in size, God could vary His angel sizes in some other way. He might start with a single angel one-half the size of the pinhead in area; create a second angel half the size of that one; a third half the size of the second; and so on.* The argument is that God can place as many angels, each half the size of the one preceding, as He wishes, without exhausting the pinhead space, as long as the total size of them all is not greater than *twice* the size of the angel we began with. The reason is this: One angel half the size of the pinhead plus ½ angel, plus ¼ angel, plus ⅛ angel, plus 1⁄16 angel . . . plus 1⁄128 angel added together will never equal two angels. Try it:

$1 + \frac{1}{2} = 1\frac{1}{2}$
$1\frac{1}{2} + \frac{1}{4} = 1\frac{3}{4}$
$1\frac{3}{4} + \frac{1}{8} = 1\frac{7}{8}$
$1\frac{7}{8} + \frac{1}{16} = 1\frac{15}{16}$
$1\frac{15}{16} + \frac{1}{32} = 1\frac{31}{32}$ etc. . . .

Note that, although the sum of the angel sizes approaches but never reaches 2 (double the original angel-size), it gets as close to 2 as one could want. The angels won't exhaust the space of a pinhead unless it is at least big enough to contain 2 of our first angels and not a hair smaller than that. Let's imagine that the pinhead is 1.9999999 original "angel-size"—pretty close to 2, you might think. God could then create 1 more tiny angel and it would fall off. To contain an *infinite* number of angels, the pinhead must be exactly equal to 2 original angel-size, nothing less. In this sense, the diminishing angels *define* the size of the pinhead.

*For the sake of the example, when I refer to a half-sized angel, I mean one that would take up half as much space on a pinhead, not one that is half as tall, half as wide, and half as thick as another angel.

What God or God's brightest interpreters had come upon was the principle of infinite or unlimited divisibility. God can put an unlimited number of angels on the head of a pin as long as each is half the size of the one before. Because the last or "littlest" angel can always be divided in half again, the potential number of angels is unlimited. Of course, the size difference between successive angels will be less and less, until, near the end, the difference is almost nonexistent. But, however crowded the pinhead, some empty space will always remain; $1/128$ "angel-size" is not much room, but it is some.

Notice that no angel will ever be zero in size. Just as the sum of an infinite number of angels defined the number 2, the trend of the vanishing angels defines the number zero. Of course, we all know that, in a simple way, zero means "nothing," but if you were to imagine even an *infinitesimally* small angel, God could create one smaller. The ability to define a finite amount by a series is called "convergence," and the numbers that the series define (in our case, 2 and 0) are called "limits."

The Greeks also thought about infinity. Like the monks, they liked to speculate on the basis of pure logic, and they had their own favorite puzzle involving infinity. Imagine a (theoretical) chase between a slow-moving tortoise and a speedy hare. If the hare starts out at some distance behind the tortoise—given a kind of handicap—at some point, it will halve the distance between them, leaving the other half yet to cover. At the next (imaginary) point, assuming the two animals keep running at the same relative speeds, the hare will have covered half the remaining distance, and, at a later point, one-half of that. As the Greeks set up this theoretical puzzler, the hare will never catch up to the tortoise, because *there will always be one more half-distance left to cover.* Another way of saying this is that there is an *infinite number* of half-distances between them. To be sure, these half-distances will become shorter and shorter—you guessed it, eventually *infinitesimally* short—but never will the hare cover the very last half-distance on this imaginary track.

The Greeks called this puzzle "Zeno's Paradox" after the philosopher who thought it up. It is a paradox, of course, because we know that a hare would easily pass a tortoise in any real-life race. The solution to the paradox lies in limits. If we measured—say, in inches—the distances between our runners at each "halfway point," we would see the distances becoming smaller and smaller—coming closer, that is, to zero. Thus, the series of distances between them (just like the series of angel-sizes) *defines* the number zero—has a *limit* of zero. Only where there is an *infinity* of half-distances can we imagine the hare, in this paradox, overtaking the tortoise.

Silly as this problem sounds, what the Greeks were trying to explain was not silly at all. They understood that, even though our brains cannot comprehend them, there are infinities all around us. In particular, they were trying to conceptualize space and time, which we think of as continuous (i.e., uninterrupted), as a collection of *points,* and this model, enhanced by calculus, gave physics, beginning with Galileo, a way to begin to describe motion and force quantitatively.

Infinity and Infinitesimals

Prior to Newton and Leibniz, infinities and infinitesimals were known, but people didn't know how to handle them mathematically. Greek geometers came upon infinitesimals (though they didn't call them that) for the first time when trying to find the ratio of the diagonal of a square to the length of any side. Those ratios turned out be nonrepeating and nonending (that is, infinitely long) decimals. The Greeks named these "irrational numbers" and learned to handle them simply by cutting the decimal off. The last digits, they reasoned, represented such very small amounts that they could be discounted.

The "discovery" of zero, brought to Western civilization by the Arabs from India around the eleventh century, added to the speculation about "infinity." "Infinitely small" came into mathematics when Descartes and others tried to find a mathematical expression for the "slope" (the "grade") of a curve produced by an algebraic function (see page 214).

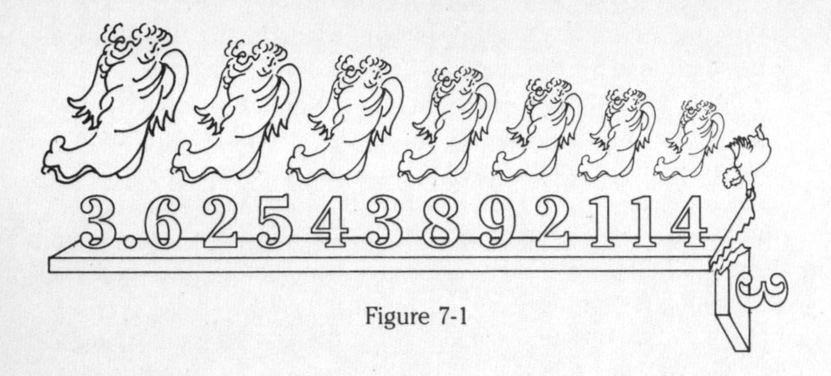

Figure 7-1

It remained for Newton and Leibniz to turn all this mathematical experience into theory. For them, infinitesimals would be treated like ordinary algebraic quantities, signified by Δ (spoken, "delta") preceding x or y or any other symbol.

Since the Δ in calculus always precedes an algebraic symbol, this means that Δ is not just any old quantity but an *increment* of that quantity, a *change* in its value. Changing values, as we shall see, is legitimate in the context of *functions;* indeed, it is the essence of functions. Typically in a calculus expression, a change in one variable, however small, produces a change in the value of the function. For example, if the function of y is given as $10x$, when x equals 2, y equals 20, and when x equals 3, y equals 30; Δx in this case is 1 (x went from 2 to 3 in this example) and Δy is 10 (y went from 20 to 30). The ratio of the change in y over the change in x is $^{10}/_{1}$ or 10.

But that only tells us the ratio of the change in that one specific numerical example. Mathematicians want to generalize the change over the entire function. Hence, they do the same calculations with x's, Δx's, y's, and Δy's. The new value of x is written as $(x + \Delta x)$, the new value of y (in our example) as $10(x + \Delta x)$. They are particularly interested in generalizing the *relationship* between the change in one variable to the change in the function. This relationship takes the form of a *ratio* between the effect and its cause—or, as they would write it, $\frac{\Delta y}{\Delta x}$.

It is only in our example that the deltas neatly cancel each other out. They do not always conveniently do so. In almost all but this simplest case, small incremental quantities remain and have to be dealt with as infinitesimals.* In calculus, the infinitesimal may be eliminated by setting it to gradually approach zero (i.e., giving it a *limit* of zero), expressed by the symbol →, as in $\Delta x \rightarrow 0$.†

Zero in calculus is not like zero in arithmetic. Calculus, as we have seen, deals with very small finite amounts. As long as Δx is only approaching zero, we can play the calculus game. When Δx gets to zero, the game, as far as calculus is concerned, is over. The change in x (or any variable, for that matter) can be made infinitesimally small, but not so small as to be nothing at all. Zero is the limit of the process, but, just like Zeno's hare, the process never brings us there. This is what makes the idea of the limit so powerful a tool in calculus. It is possible to have Δx's and Δy's in the algebraic manipulations until the last moment, and then we can eliminate them (the *limit-going* process) by imagining them slowly shrinking to zero.

All of this seems bizarre to those of us who never graduated from elementary to more advanced algebra. At the beginning of algebra, when we were first introduced to symbols like x, we were told to "solve for x," meaning we were to discover the hidden number that x represented. Most of us thought of x as standing for "answer." In advanced algebra, functions, and calculus, however, x, y, and the other "unknowns" do not stand for unique number-equivalents, but are, rather, *variables* in fixed relationships with other *variables.*

If I were to change the mathematical curriculum in any one way, I would ask teachers to make clear the distinction sketched out here between x as the "answer" and x as a variable. Before going on to calculus, students ought to be thoroughly familiar with the idea of variables and their place in the mathematics of *functions.* The tragedy for most of us is that we never saw that

*So central are infinitesimals to calculus, in fact, that the name for calculus in German is *Infinitisemalrechnung,* meaning "calculation involving infinitesimals."

†Mathematicians use the symbol $\frac{dy}{dx}$ in finding the limit when Δx approaches zero.

functions provide a bridge between algebra and geometry. In most places, algebra and geometry are still taught as separate subjects, the one (algebra) as a manipulation of x's and y's, the other (geometry) as a collection of impossible-to-memorize proofs. What I realize now (and wish I had known then) is that many of the mysteries of calculus are cleared up by thinking geometrically about functional relationships.

Calculus: Where Algebra Meets Geometry

A function is a formula in which specific numerical values given one unknown produce specific numerical values in another. For the past 350 years, however, the algebraic function has been thought about in terms of geometrical figures. Descartes invented the format: an x axis in the horizontal direction to show all the numbers used, and a y axis in the vertical direction to show all the numbers yielded. When points are placed on the graph so that directly below each one, on the x axis, is a number used, and directly across, on the y axis, is the corresponding number yielded, a picture is formed that is unique to that function. Usually these "pictures" are lines or curves. The more complex the formula, the more complex the picture. Descartes called this a "system of coordinates," and we honor him still today by calling these "Cartesian coordinates."

You start with a function such as $y = 2x + 3$. Then you choose some values for x and calculate the equivalent values (given the equation) for y.

x	y
0	3
1	5
2	7
3	9
4	11
5	13

and so on.

Then you *plot* the numbers as points along the coordinate system and connect the points; you have a line. (See figure 7-2.)

When higher powers are involved, such as in the function $y = 2x^2 + 3$, that "line" will be curved. (See figure 7-3.)

x	y
0	3
1	5
2	11
3	21
4	35
5	53

You can already see, just from the table, that in this function y is increasing at a much faster rate than in the first. And when you plot the points, the resulting connection is a curve.

The Cartesian system allows you to draw the function so that you can *see* the resulting slope of the line, or curve. But one problem remained that Descartes, despite his genius, was never able to figure out: How to go from the graphic or geometric solution back to the algebra? How to express algebraically the slope of the curve? Here is where all the elements of calculus come together.

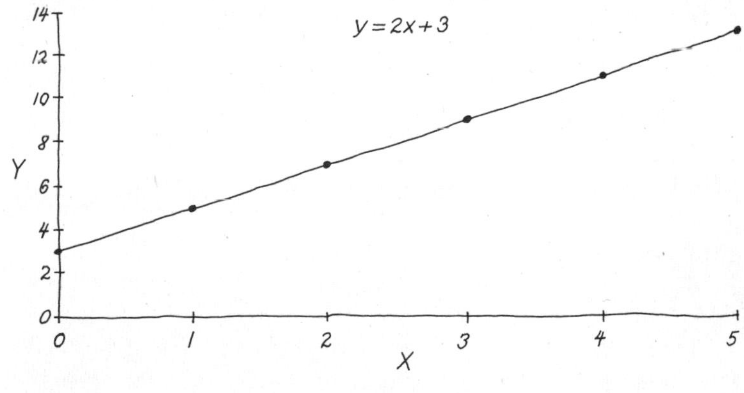

Figure 7-2

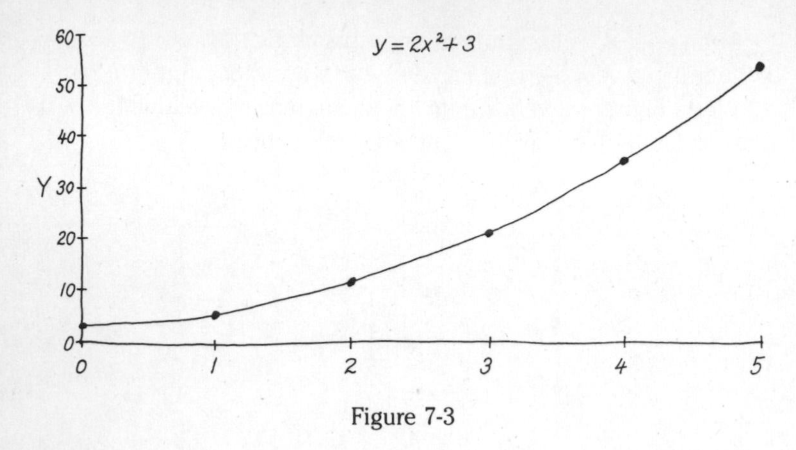

Figure 7-3

Slope

What is a slope? Highway signs indicate slope as "grade"; when the sign says "6% grade," it means you rise (or fall) in altitude by 6 feet for every 100 feet you travel along the road. So much for the slope of a line. What about the slope of a curve? A line has the same slope at every point; a curve has a different slope at every point. (See figures 7-2 and 7-3.) Mathematicians call the slope of a curve at a particular point on that curve a "tangent." Whether we are talking about the slope of a line or the slope of a curve, the slope is defined as how much *y*—the altitude—increases (or decreases) as *x*—the horizontal distance—increases (or decreases) by some amount.

The hypotenuse (dotted line) of the right triangle with Δx and Δy as its sides is a straight line, not too different from the actual path (the solid line). (See figure 7-4.) But the hypotenuse is of course only an approximation of the slope of the curve at a particular point. The trick is to shrink the size of the triangle—that is, decrease the side lengths toward some "zero" length—so that the hypotenuse is virtually (though it will never be exactly) the curve segment itself. (See figures 7-5 through 7-9.)

This is where the discussion of angels and limits becomes so pertinent. Now, instead of imaginary angels, we have a real picture with real measurements. We can shrink our triangle, just as

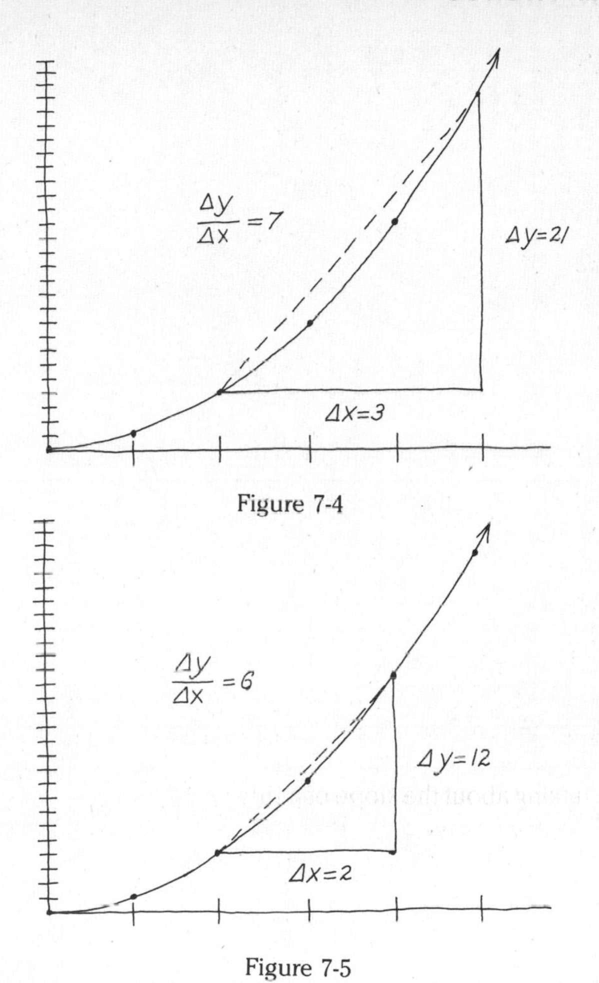

Figure 7-4

Figure 7-5

God shrunk the angels, and measure, each step of the way, the slope of our hypotenuse. As we get smaller and smaller, our measurements will come closer and closer to one specific number. In the sequence of figures below, that number is 4. Our measurement will never reach that number, (that is, we will never have *nothing* to measure), but we can see that, if we did every possible measurement, the "last" one—not that there is a "last" one we can reach, only a "last" one we can aim for—would reveal the

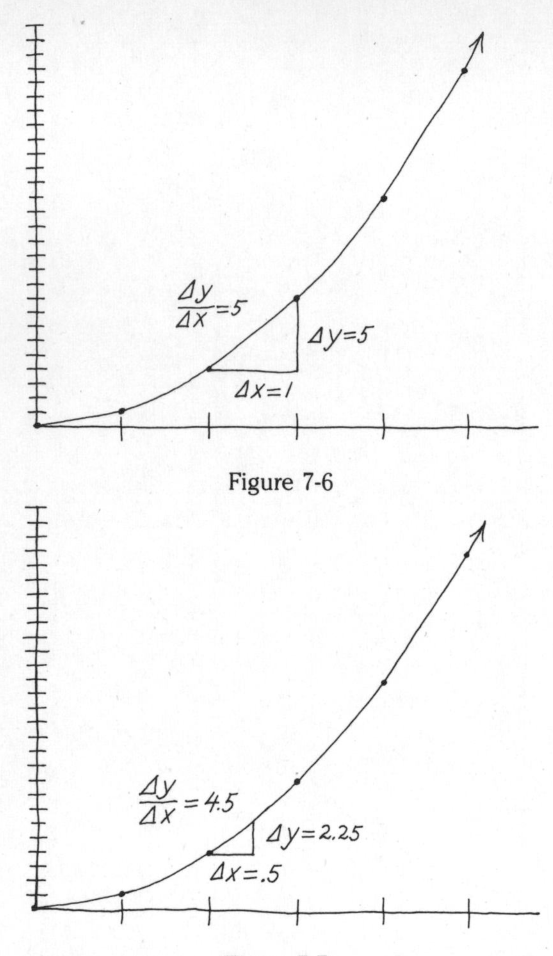

Figure 7-6

Figure 7-7

correct slope. Now it is our measurements (not an imaginary sum of an infinite number of angels) that define the answer.

There are still some calculus and physics classes in which students use rulers and pencils to find the slope at a point on a curve. Recent advances in technology let students do this procedure on a graphing calculator. However, since most scientists and mathematicians don't have either the time or the artistic ability to find the slope in this way, they use the precision of

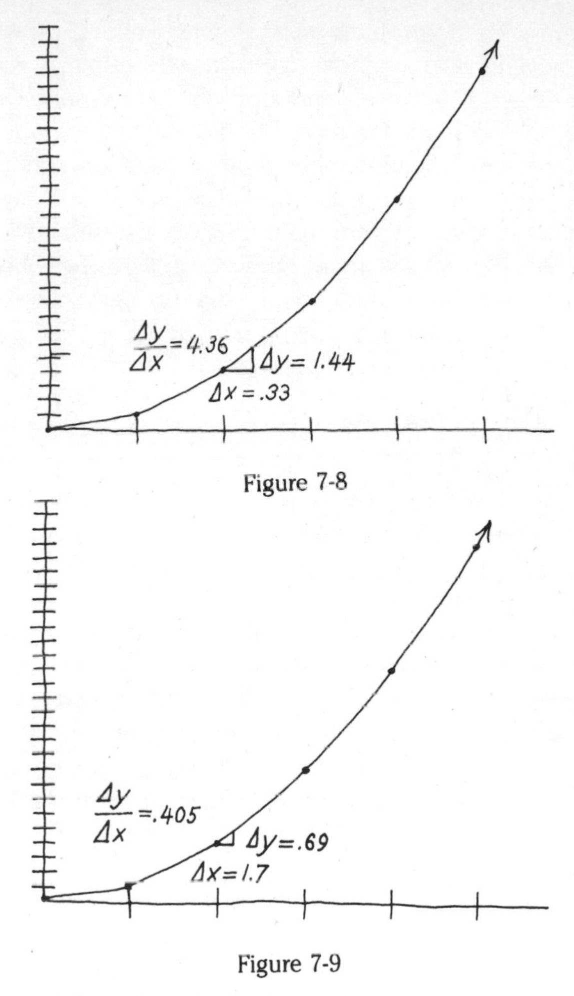

Figure 7-8

Figure 7-9

algebra and the rules of calculus to find the exact measurement of slopes.

Calculus and Its Applications

What are the uses of calculus in solving real-world problems? Why does it inspire so much awe? To answer that question, we can begin with the problem Newton was working on when he invented calculus—namely, a quantitative description of motion.

What do we know about motion? Objects in motion change position at every instant of time. But motion is continuous. How can we analyze something happening continuously when all we can measure are points in space and instants in time? Newton did it by reducing to nearly zero the time intervals during which the object changes position. Nearly zero is as close as he (or we) can get to instantaneous change. Well, what is "nearly zero"? How about "infinitesimally small"? And how do we get to "infinitesimally small"? By diminishing the interval.

DIMINISHING INTERVALS

To get at any event as it is happening is not easy. We can measure what we want to measure "before" and "after," but how do we put "during" into a box? "During" can be roughly considered as the space or time or distance or gap—mathematicians call all of these "intervals"—between "before" and "after." The nearer "before" is to "after," the more precisely we can measure "during." Any moment or half-moment or quarter-moment or one-hundred-twenty-eighth of a moment before something happens is still "before"; any moment or half-moment or quarter-moment or one-hundred-twenty-eighth of a moment after something happens is "after." The closer we can get to the "during" by diminishing the interval between "before" and "after," the more likely we are to capture it.

Think of movies, which are fast-moving still pictures flipping over fast enough to fool our brains into believing the pictures are in flux; or animation, based on the same principle. Every movie frame represents a thirty-second of a second. But to capture high-speed events, such as a single bullet shattering a piece of glass, the camera speed has to be increased a hundredfold. This means each frame corresponds to a thirty-two-hundredth of a second. The more frames, the shorter the interval, the closer we come to capturing the instant on film.

The trick is to imagine an infinitely small interval, a curve of "zero length," a line segment barely larger than a point, and to place this imaginary construct upon the unmeasurable or uncountable event.

MAXIMUMS AND MINIMUMS

"Limits" can be helpful in another sense. In real life, we may want to know at what point a process, a manufacturing plan, or a government policy reaches its optimum usefulness; or, to say it differently, the moment just before there are diminishing returns. If we have a system for measuring continuous and instantaneous change, we can find out pretty easily at which point in the process *no change* is taking place. For an increase in pay per hour, for example, some number of employees will be willing to work extra hours or an unpopular shift. But after some number of extra hours, many of the employees will value the money less and will prefer more leisure time to more work.

In modern business planning, past experience with seasoned workers will be plotted on a graph that may, in turn, be displayed on a computer screen, so the slope of the curve can be observed. The point at which the curve is no longer going up but not yet going down is called a "maximum" (figure 7-10). The maximum is the point on the curve where its tangent is horizontal, its slope

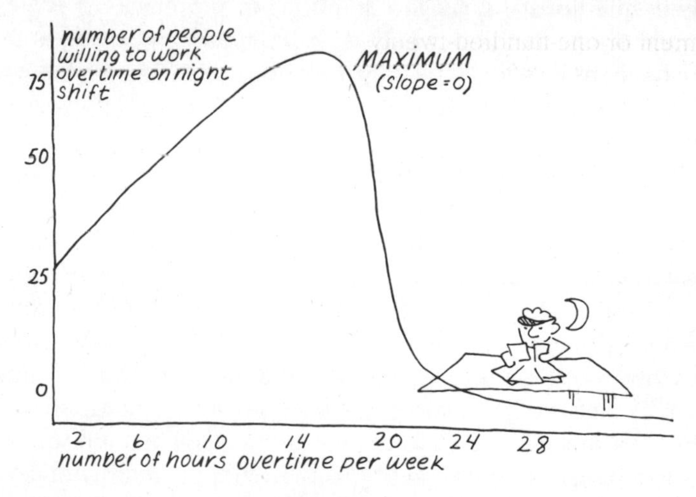

Figure 7-10

zero. For the business planner, this represents the minimum wage to keep employees working full- or overtime.

Both maximums and minimums have zero slope. The maximum is the point after which the curve begins to turn downward. The minimum is the point at which the curve begins to turn upward.

Or, to take another example, some cutback in expenses in running a retail store will not affect revenue. But at a certain point, such as closing the store on Sundays or not repairing the neon sign, a change in policy will discourage so many customers that there will be a change in the direction of gross sales. In fact, this pattern is very common in many real-world business situations. One hits a peak, after which there is a flattening out, a plateauing of the effect. These effects can be dealt with mathematically by means of techniques that trace change as it occurs. If we know the slope of a curve and its high and low points, we have valuable information about a system: its rate of variability and the limits to that variability.

After making a mathematical model of the situation, calculus permits professionals to find maximums and minimums, points at which the slope is zero. Such techniques can answer many urgent questions and suggest solutions to problems. As professionals, we do not have to do these calculations ourselves. As long as we understand how they are done, we can assign them to others.

SUMMING SMALL CHANGES

Suppose you had an attic or a dormer room with a curved, uneven ceiling, and you wanted to know the exact area of one of its walls. How would you measure the area of a rectangle with a curve instead of a straight edge along one of its sides? Measuring the area of a rectangle is easy. Rectangles are closed and regular, and their areas can be calculated by the straightforward formula "length times width." It is quite another task to measure area when the shape is curved and nonuniform. One method worked out two thousand years ago imposes an infinite number of measurable rectangles on the area under a curve until the total area of the rectangles approaches the area under the curve. (See figures 7-11 and 7-12.)

Compare figure 7-11 to figure 7-12. The curves in both cases are identical, but the rectangles drawn within the one are half as wide as the rectangles drawn within the other. The technique is to create figures that can be measured, in this case rectangles, and place them on an unmeasurable space. Reducing the little triangular chips that extend above the rectangles allows an ap-

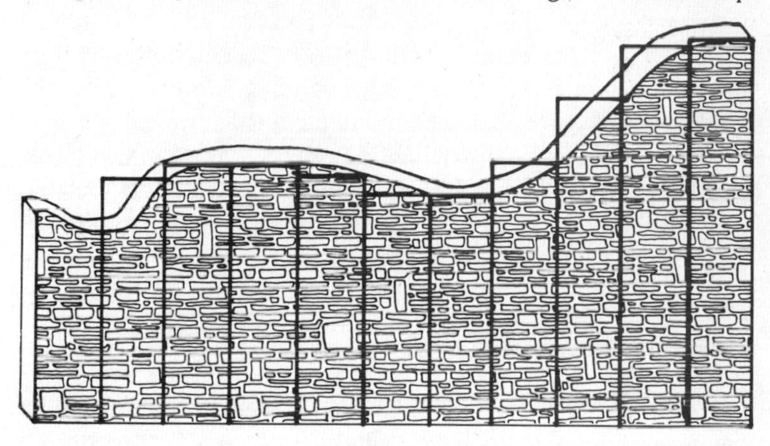

Figure 7-11

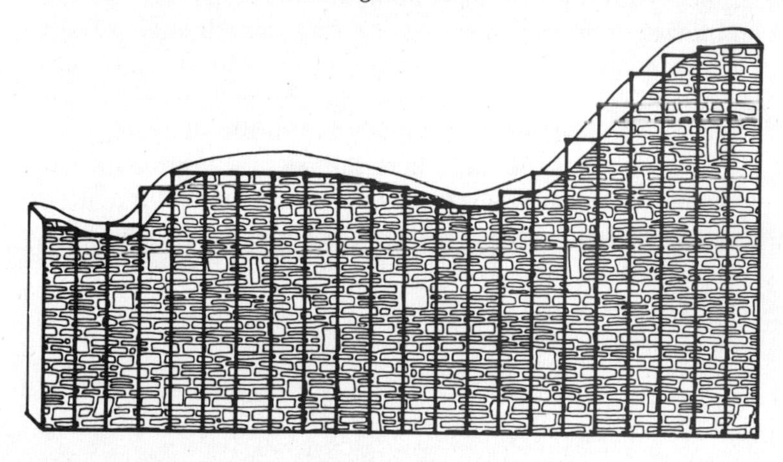

Figure 7-12

proximation to be made. Where there are more rectangles (figure 7-12), the excess triangles are smaller and the approximation is more exact. Eventually, with even narrower and more numerous rectangles, the difference between the area we *can* measure and the area we *want* to measure is so small as to be insignificant. Calculus permits this: to figure out the original function that produced the curves and algebraically calculate its area.

Once we can capture a relationship between variables by calculating the slope at every point along a curve, and once we can measure the accumulated effect of all those relationships by finding the area under the curve, we can begin to examine some of the perpetually changing phenomena of the physical universe. This is the value and the delight of calculus: It allows us to describe not only events in flux, but events whose rate of change is itself in flux. In a world where things are changing, often at different rates, calculus is indispensable.

Conclusion

Where does one go from here? If calculus were only a method for solving engineering problems, I would not go any further. But it is far more. It is a way to think about relationships among things, and a way of ranking their impacts. And since it offers a way to cope with independent and interdependent things happening at the same time, a person trained in calculus will be able to use a more realistic mathematical model to study the situation.

This is the good news: Out there, just over the horizon for many of us, is a mathematical system that can enhance our ability to think about and increase our mastery of complex issues. Now for the bad news: There is probably no short cut to mastery.* We can

*The concepts discussed in this chapter—infinity, infinitesimal, limit, and convergence—are the conceptual challenges of calculus; the solving of calculus problems is, I am told, relatively easy, as long as a student's algebra skills are strong. Many students do not let it bother them that they don't fully understand what they are doing. They simply follow the rules, accumulate experience in doing calculus, and wait for the fundamental concepts to emerge.

talk about calculus only so long. At some point, as when we study a foreign language, we will have to master the vocabulary and the rules of grammar, do the exercises, and actually translate.

Getting from here to there seems easier now, at least for me. It was my reflections on the meaning of infinity—what infinitely many slices of infinitely small portions will do—that set me thinking about divisibility, intervals, limits, and slopes as I had never thought about them before. As my appreciation of calculus increased, my fear of it decreased, and when I substituted words that made sense to me for alien notation, I began to incorporate calculus into my own repertoire of thought. Eventually, for me, the subject became a coherent set of ideas, not just another branch of mathematics.

Thinking about infinity need not have been the starting point. I might have asked myself how the speedometer provides us with our speed at every instant in time; about the area under a curved line on a graph,* or about the origin of *pi.* Somewhere out there for each of us is a metaphor that will make sensible the shapes and squiggles that give us pause when we open a calculus textbook. And to open our minds to the possibilities of calculus, in my view, any metaphor will do.

Notes

This chapter, in its first incarnation, was aided enormously by conversations with my friends and colleagues who really understand (and teach) calculus. Susan Auslander, Joel Schneider, Peter Hilton, and Carlos Stern were particularly helpful in steering me away from pitfalls and encouraging me to pursue my own line of thinking. In the revised edition, Carl Tomizuka, a physicist, and Gary Horne, a teacher of mathematics, have been of invalu-

*The speedometer does continuous a differential-calculus calculations. One calculus textbook (Tom M. Apostol, *Calculus,* Vol. I, 2nd ed. New York: John Wiley & Sons, Inc., 1967) actually starts with questions about area, instead of questions about slope.

able assistance to me. I am, however, as before, responsible for the elaboration of the "scholastics'" approach.

1. Sylvanus P. Thompson, *Calculus Made Easy* (New York: Saint Martin's Press, 1910 and 1986); Eli Pine, *How to Enjoy Calculus* (Hasbrouck Heights, N.J.: Steinlitz-Hammacher, 1984).

8 Overcoming Math Anxiety

The best therapy for emotional blocks to math is the realization that the human race took centuries or millennia to see through the mist of difficulties and paradoxes which instructors now invite us to solve in a few minutes.
—Lancelot Hogben

Quite probably, hidden in our minds in recesses that are not easy to expose, is much of the math we need to enhance our occupational choices and enrich our personal lives. In digging out these dimly remembered procedures, we may notice some of the mathematical ideas that escaped us the first time around (as we saw in chapter 6) and be ready for more advanced and stimulating math (like calculus). To get at these recesses, I believe, we need only some knowledge about our own minds and how they work, about our temperament and fear and how these intrude on good thinking, and some simple exercising of our rusty pipes. If we can control our anxiety by recognizing its symptoms and coping with them, we can go far.

How far do we want to go? Not far enough to become engineers or mathematicians, perhaps. But surely far enough so that our fear of math no longer makes decisions for us.

What follows is no surefire cure for math anxiety, but a set of suggestions. Some of these techniques involve professionals. Math teachers or counselors or both may be necessary at the outset. Other techniques can be developed by oneself and for oneself. Some concrete materials, readily and cheaply available, will help elucidate certain numerical and geometrical relationships. Still other concepts may require some reading in the history of mathematics or in the psychology of learning, to provide some theoretical underpinning for what we are doing. All require that we take some initiative in our learning process. As W. W. Sawyer, a mathematician and an inspired teacher of mathematics, puts it:

> In discovering something for ourselves, we have a sense of freedom and conquest. In memorizing something that another person tells us and that we do not understand, we are slaves.

The first step in overcoming math anxiety is to *take charge* of our math learning, to stop being intimidated both by our own lack confidence and by hallowed traditions in the math classroom that keep us from feeling *good* about ourselves. Long ago, Sandra Davis, working to allay students' math avoidance, came up with the now famous "Math Anxiety Bill of Rights":

Math Anxiety Bill of Rights
*By Sandra L. Davis**

I have the right to learn at my own pace and not feel put down or stupid if I'm slower than someone else.
I have the right to ask whatever questions I have.
I have the right to need extra help.
I have the right to ask a teacher or TA for help.
I have the right to say I don't understand.
I have the right not to understand.

*Distributed by the author at math teaching and learning workshops.

I have the right to feel good about myself regardless of my abilities in math.
I have the right not to base my self-worth on my math skills.
I have the right to view myself as capable of learning math.
I have the right to evaluate my math instructors and how they teach.
I have the right to relax.
I have the right to be treated as a competent adult.
I have the right to dislike math.
I have the right to define success in my own terms.

Math Anxiety—The Clinical Approach

"My name is Carlene. I hate math intensely." So begins the first go-around at a no-charge, no-credit math-anxiety session held at the University of Arizona. Carlene is one of a dozen or so undergraduates and adults gathered to explore their math strengths and weaknesses, many for the first time in years. The sessions are being held in two neighboring classrooms, the one a comfortable seminar-style room which, the instructors hope, will stimulate free-wheeling personal discussion, even self-revelations. At times the group moves to the "math room," a more standard classroom, where the instructors and students can take their work to the chalkboard.

"I don't hate math," says Terry. "But I just find it real hard." Scott hasn't had a math course since 1975. He is in construction work and finds that he does the "concrete stuff" pretty well, but "when you throw in the abstract symbols, like *a* and *b*, I lose control." Lucia says she gets frustrated when the numbers and the letters are mixed. "I don't think it's right," she says, as if there were some divine commandment: "Thou shalt not mix numbers and letters." Everyone laughs.

Pamela Reavis and Gary Horne, who, with graduate student Mary Ellen Hunt, have organized the class, are experienced high-school math teachers. Both majored in math, which the assem-

Register now for

Math Anxiety Clinic

π $\sqrt{\ }$ Σ $\int$

Did you hate your math classes?
Does math make you feel dumb?
Would you like to feel better in six weeks?

Don't delay! In six weeks, you could look at math in a whole new way!

Classes are every Tuesday and Thursday
from 7:00 to 8:30pm
starting Oct. 6 and ending Nov. 12
in Family Consumer Resources Bldg.
Room 217
Cost is only $7.50 for a xerox packet

For information and sign up, call Mary Ellen Hunt
621-7339

Sponsored by the WISE office. Gary Horne, Pam Reavis, instructors.

bled math-anxious adults quite obviously find bizarre (although they are too polite to say so). Gary tells them that, from his own experience as a math major and teacher, he has observed that good math students ask a lot of questions. The purpose of the twelve sessions that are ahead, he explains, is to get these timid souls to engage, asking whatever questions they need to have answered so that they are never left behind. The course has no particular syllabus. Each session will begin, under Mary Ellen's baton, with "unfinished business," math material and emotional baggage left unresolved at the end of the last one. There will be some motivation and anxiety-reducing activities in every session, but math instruction at a level that interests the group, too. "We want this to be a good experience for you," says Pamela, "maybe the first good experience you have ever had in math. We'll do some math, we'll talk about your feelings, and then we'll have some food." Indeed, lemonade and a plate of cake and cookies await.

Gary starts with a slide show designed by inventor Paul MacCready (see chapter 5). MacCready's work, as we have already seen, is intended to stimulate creativity in problem solving. But first you have to concentrate on some detail. "Count the *yellow* dots in this picture," instructs Gary, flashing a slide with yellow, green, blue, and red dots. The class obeys. He puts the slide away. "Now, how many *red* dots were there?" No one knows. This illustrates how we see only what we are supposed to see, and do not notice what's outside the frame. Next is the nine-dot display we explored on page 158.

Finally, Gary flashes a MacCready Count the F's exercise on the screen.*

FINISHED FILES ARE THE RESULT OF
YEARS OF SCIENTIFIC STUDY COMBINED
WITH THE EXPERIENCE OF MANY YEARS

*Of course the F's weren't highlighted when presented to the class.

Almost no one remembers to count the F's in the little words, like "*of.*" The students are slowly beginning to grasp the purpose of these exercises—to give them some insight as to how much or how little detail they are willing to pay attention to. He has gotten them to *think* about *thinking*. And that's what this Math Anxiety Clinic is going to build on.

Soon the group make their way to the "math room." They do two problems, one at their desks alone, the second in conversation with a neighbor. So far, this is like a math class, but with an important difference. Once the group has finished struggling with a problem, they engage in extended conversation about what they did and what they were *feeling* as they worked the problem. Lucia was inclined to give up on the first one, she says. She was sure she had the "wrong answer," and she had learned in her previous math classes that the "right answer" was all that mattered. But not in this class. All the "wrong answers" are collected; people's logic is taken seriously. Val, who has no idea where he pulled his (correct) answer from, is encouraged, instead of punished, for "not showing your work," as he would have been in school. "Try to reconstruct your thinking," Pam urges. And the rest of the class assists.

Thus, in the space of the first ninety minutes, the class has had a glimpse of all the various ways people approach a "simple" problem, and all the creativity involved. Indeed, this is going to be a very creative group. That is obvious, after the first class, to everyone. Yet, earlier that evening, none of the participants would have thought of himself or herself as being "creative" in math, or that math involved creative thinking and not just rote memory. They have crossed the first hurdle to overcoming their math anxiety. They have stopped disparaging themselves.

Two weeks later, the group has made some progress with fractions, percents, and word problems. However, Pam, Gary, and Mary Ellen are convinced that anxiety and resistance are still a problem for them. So they start in again, this time with a fill-in-the-blanks exercise to take a measure of where they are:

When I make a math mistake, I ______.
When I'm embarrassed about doing math, I ______.
When I see a problem I can't do, I ______.
If I could do math, I would ______.
One thing I like about doing math is ______.
Doing math makes me feel ______.

Discussion reveals that members of the group are still judging themselves by how well and how quickly they do their math. "If I see a problem and know what I'm doing, I feel smart," says Lisa. "But if I don't, I feel negative." Carlene still doesn't know why she needs to learn this stuff. "In real life," she says, "if I needed to know the actual size of a room, I could measure it. If I had to wallpaper an area, I'd just buy some and take back what's left." They are working hard, but resenting how difficult it is to relearn math. And several are still not sure why they have to.

Pam tells them that doing math is like keeping in shape so that, "when you need the skills, you'll have them ready." Instead of fighting the prevailing malaise, Gary compares their feelings to his sense of powerlessness in the face of English literature. Lucia is the most poignant. "I know what I'm missing. I want so to be able to read about physics and astronomy, but math stands in the way."

It takes only a casual observer to see this is no ordinary math class. Even when the students are in the "math room" learning new techniques and doing problems, the bonding among participants is palpable. The high instructor/student ratio pays off. The three instructors parry with one another and with the students. When the class breaks for problem solving in small groups, an instructor will join a trio, providing support and, where necessary, some hints.

Group Process

Some of the activity in these math-anxiety classes will be recognized by counselors and psychologists as no more than good group therapy. Others will see that its essence is in collaborative learning. In place of the kind of competition (math-anxiety victims often call it "assault") that many adults associate with their earliest experiences with mathematics, Horne, Reavis, Hunt, and Arem (see page 244) are all determined to show their students how much they can learn from one another, and how another person's capability need not be a threat to their own.

In the original math-anxiety workshops established by our team at Wesleyan in the 1970s, Jean Smith, our mathematics instructor, and Bonnie Donady, our counselor, would begin their sessions by sharing math autobiographies. Out of these discussions would come a survey of what caused the group members to decide to overcome their handicap. The group meetings did not follow a rigid schedule. Donady and Smith assigned "psychological homework," having the group members focus on the feelings aroused when they worked on math. The first assignment was to observe how they treated themselves when they encountered difficulty. Did they blame themselves? Castigate themselves? Or were they patient and self-reassuring? Another assignment was to watch how they handled frustrations when doing math in comparison (usually in contrast) with how they handled difficulty in nonmathematical situations. Eventually, mathematics was introduced into the group session, and the members were urged to monitor their own and others' reactions as they moved from a meeting format into a class format. As soon as anyone began to fall into a pattern of "succumb and surrender," the group leaders, if not the group itself, intervened.

If the problem for many people in learning math is translating something they do not understand into language that makes sense to them, the presence of other people much like themselves will help immeasurably. In the course of a discussion about fractions or ratios or tipping, someone in the group may use a term a math instructor might not have thought to employ in such a connection. This term will raise the curtain, and the issue will make sense to someone else. I have seen it happen when, in a group of adult women, one complained that she had always had trouble turning feet into inches and inches into feet because she never knew whether the new amount should be larger than the old one. She was helped on the spot by someone else in the group, who suggested that before she undertook any measurement computation of this sort she should note whether she was going "from many to fewer" or "from fewer to many." "Many to fewer, fewer to many," the first woman kept saying to herself. She understood, possibly for the first time, how to cope with her disability.

I learn math via a process I can only describe as dialectical. Since the usual presentation in a math text often makes no sense to me, I try to write down in prose what I think does make sense. Then I interview mathematicians or people who know more math than I, and ask them whether what I have written is correct. Sometimes it is and sometimes it isn't; but almost always I am told my question seemed to come at the issue sideways—sideways from their perspective, of course, not from mine. Trying to write out something as it would make sense to you is one way to enter into that dialectical process. It is by no means efficient, in terms of time alone. But if it works, it is the only way to begin.

There is no one right way to run a math-therapy or math-desensitization group. Nor does the mathematics discussed need to be as elementary as in the group just described. Generally it is advisable to have a psychologist and a math instructor share responsibility, so that errors in logical thinking and mistakes in computation can be cleared up immediately. Most important is to eliminate anxiety-producing experiences: no tests, of course, no pressure to get right answers, no competition with classmates, no put-downs. This reversing of the usual mathematical experience may finally give the learner one positive math experience that will go far toward reducing anxiety.

Math and Math-Anxiety Reduction Combined

Thanks to the widespread dissemination of techniques for reducing math anxiety, math instructors (especially those teaching courses geared to adults) are finding numerous ways to integrate some attention to *feelings* in their regular math offerings. This is particularly valuable since so many victims of math anxiety—particularly men—are skittish about acknowledging their fears, or about signing up for math-anxiety workshops generally attached to or sponsored by women's programs. One model of how to do this kind of integration is provided by Frances Rosamond, a math instructor with a Ph.D. in math education, and a self-taught math-anxiety counselor, who has been teaching "scared adults" for fifteen years, first in a ten-week summer-school program that she established at Cornell University in 1977 and, since 1986, at National University near San Diego.

On the day when I arrive to observe her course, I am ushered into a storefront minicampus of National University, an institution wholly dedicated to the needs of the working adult. Thirty-eight adults are trying to master some of the mathematics that has eluded them in the past. The course is standard Algebra I, taught in three intensive five-hour after-work sessions per week, and required by National for graduation. But the instructor and her approach to her students are anything but standard. Unlike Donady and Smith's, the course does not begin with "math autobiographies," nor does it focus explicitly on students' psychological needs. These students don't have the luxury (they think) of *exploring* mathematics. They have a course to pass and a degree to earn so that they can qualify for a better job and get on with their lives. Nor are they inclined to deal directly with their angers and fears. Math for them is a *subject,* not a relationship between themselves and the discipline. But Rosamond thinks otherwise. To protect them from panic, she allows them to use their calculators in class and to have at all times a personal "box" of formulas, hints, and math facts that they can carry around and refer to—even on tests. In exchange, she insists that they keep a journal in which they freely "vent" what they're feeling about her class and about mathematics more generally.

Sometime after the first test, when she has already gotten to know her students from their journals, Rosamond sets aside an hour of classtime for a discussion of feelings. At first, some are resentful. After all, isn't this a waste of the time they could spend reviewing the test? But soon resistance gives way to curiosity about themselves and about each other. Rosamond is patient.

"People keep asking me why I am taking this class," one student begins. "Like me, they all hated math in high school. But I am doing well, better than I thought I would. Either this is a fluke, and I'm going to start doing worse very soon, or those grades I got in high school weren't really 'me.' "

"I'm not interested in learning math by rote any more," reveals another. "I want to see how the parts fit into the whole." The class agrees that they are generally more aggressive in tackling the subject and more patient with themselves now than they were in high school. But the old "scars" still hurt. Rosamond wants her students to work at the blackboard not to punish them but so she can see their work. They resist, however. "I used to feel *assaulted* by other people when I went to the board," one remembers. The instructor compromises. They will henceforth only go to the board in groups.

Although hers is not a formal math-anxiety course, Rosamond brings her considerable experience with math-anxiety victims to bear on the way the class is structured. Students work in pairs. There are no "dumb" questions. Discussion ranges from how to do a problem to "What is calculus, anyway? I've heard about it all my life, but no one every tells me what it is." As chair of National University's Math Department, Rosamond can guarantee what sensitive instructors in other settings can't always—namely, that, in the math courses that follow Algebra I, students will be allowed to bring their calculators and their "boxes" with them to exams. Her position is this: In the real world of mathematics at work, her students will have all the math tools they need to do the job—calculators for arithmetic calculations, computer software like Derive (which does algebraic manipulations), and log and conversion tables, the works. Why should they have to struggle with rote learning they will never use again?

The discussion comes to a close. The students turn back to the problems they missed on the test. But the ice has been broken, trust has been established, and the instructor has every reason to believe that the group has bonded and will help her and one another over the rough spots in the three weeks of class that still remain.

Programs for Women

Many people, particularly women, need more than occasional discussion to keep them going, more even than one positive experience learning math, to overcome their fears and feelings about the subject. Kathryn Brooks, a women's-studies specialist and counselor, has been running math-anxiety courses for eighteen years, first at California State College at Fresno, later at the University of New Mexico, and currently at the University of Utah. Her strategy is to set aside eight weeks for talking about "math messages," where they come from, and how they have affected her students' lives. In addition to talk, her students keep a "math record" during the course, detailing their reactions to her program, to the mathematics they have to deal with in their day-to-day lives, and to their slow recovery from previous trauma. Her purpose: to give students back *control* over their math learning. To accomplish this, she mixes "math detox" with assertiveness training and what she calls "class-management techniques." These techniques are meant to give her students the means to get what they need from formal instruction.*

When she left New Mexico for Utah in 1989, Brooks teamed up with Ashley DuLac, a mathematics graduate student, whose interest in math anxiety has now virtually replaced her desire to pursue a career in pure mathematics. Because of funding pressure, Brooks and DuLac had to compress their math-anxiety program into six intensive sessions. So, to start the process of personal

*"Reducing Math Anxiety" continues at the University of New Mexico as an offering of the Women's Studies Department. It is currently taught by Bev Herbert.

"detox" right away, the instructors hand out a list of "math messages"—in the form of "Do You Believe?"s—on the very first day of class:

Do You Believe?
Courtesy of Ashley DuLac and Kathryn Brooks[1]

Men are better in math than women?
Math requires logic, not intuition?
Mathematicians are smarter than most people?
Some people have a math mind, others don't?
Math is rigid?
Mathematicians do problems quickly, in their heads?
Mathematicians rarely make mistakes?
There are magic keys to doing math?
It's bad to count on your fingers?
Math is not creative?
Mathematicians are eccentric?
Math requires lots of memorization?
Most people don't have to know math for daily living?
In doing math, it's important to get the answer exactly right?
Math problems are done by working intensely until the problem is solved?
There is always one best way to do a math problem?
You must always know just how you got the answer?
Math is fun?

Participants at Utah and elsewhere are surprised to find out how few of these common beliefs about mathematics are true. Math isn't done instantly. Successful students of math will read through a math problem many times and only then settle on a course of action. There is rarely just one right way to get an answer. In fact, the more interesting mathematics comes from thinking about how *else* one could have solved a problem. And as for mistakes, once students get over the trauma of making them, mistakes can be "interesting." One by one, the "messages" are exorcised, but, even more important, their impact on these students' lives comes fully into view.

The Math Autobiography

The classic dredging-up technique for recalling past feelings about mathematics, first conceived by Bonnie Donady, is the "math autobiography." Orally or in writing, math-anxiety victims are invited to re-experience what went wrong. It doesn't really matter what form is used. As soon as people come back to math, their early negative experiences rapidly surface.

Jean Smith employs this math autobiography form to get discussion going:

1. My main goal in taking this course is ______________________________
____
2. The last math course I took was ______________________________
____
3. An early experience I remember in math class was ______________
____
4. One math teacher I remember is ______________________________
____
5. I feel ______________ was the hardest for me to learn and ______________ was the easiest.
6. I think I learned my present attitude toward math when ________
____
7. To improve my math attitude, I expect to do the following for myself ______________________________________
____
8. To improve my performance in math, I expect to do the following for myself ______________________________
____

The math autobiography calls up sadness, quite as much as prior frustration. Smith remembers one woman in particular from the course for adults at Wesleyan:

> I can never remember liking math. I had a problem, only diagnosed when it turned out my son had it, too, that caused me to

> reverse numbers. When I multiplied 7 × 3, I knew the answer was 21, but I wrote down 12. Learning disabilities were unknown at the time. The school told my parents I was socially immature. But while I was having trouble with arithmetic, my sister was talking trigonometry to my father. It led to terrible feelings about math.

Brooks and DuLac find a similar recollection:

> I attended a small rural school where algebra was not offered until ninth grade. I had spent my entire freshman year as an exchange student, learning a foreign language, and had missed out on the opportunity to take algebra with the rest of my small class. So I began algebra as a sophomore, one year behind the rest. From the beginning, the tone was set by the seating arrangements. The boys were grouped on one side, the girls on the other. Without exception, none of the girls was considered bright. When, on the first exam in algebra, I received the second-best score in class—117 out of 120—my teacher asked me if I had cheated. From there on, it was all downhill.

Since so many of the people who seek out math-anxiety courses are adults returning to school or trying to re-enter the work force, their children's experience throws light on their own.

> Recently, I attended parents' night at school. The physics teacher asked me how my daughter liked his class. When I told him she was having a little trouble with the physics concepts, he said, 'That's not unusual for a girl. They have a real hard time picking up these ideas. She'll catch on in time. Girls just take longer.' I must admit the message was double-edged—there's a certain safety in knowing 'girls take longer' but there's also condescension in that teacher's attitude.

Though hostile teachers and a negative "classroom culture" often surface, the point of the airing of the math autobiography is not to locate blame so much as to recognize the sources of the problem for a particular person. Math fear often turns out to be thinly veiled resentment. After all, the victims were young when they first were embarrassed or intimidated in math class. It is inappropriate for children, particularly girl children, to demonstrate anger. Hence, anger frequently becomes transmuted into fear. But it doesn't take long in carefully structured math-anxiety discussion for the connection to be made. "I'll show you," one woman finally blurted out to the imaginary fifth-grade teacher she was trying to exorcise from her memory. "I *won't* do math." She was as surprised as her instructors to discover that it wasn't anxiety but hostility that had fueled her resistance to math.

Talking About Math

People who don't like math don't like to talk about math. Part of their avoidance mechanism is to pretend that it does not exist. But math does not go away. People need it at work, in calculating percentages, in dining out, and in handling money. A math clinic is designed to integrate talking about math into the learning process.

Most educators know that feelings influence learning, but few take time out of class to deal with feelings. They may think that such a discussion is a distraction from the work at hand. One school in New York City experimented with a math clinic and had quite a different experience. High-school students were taken out of their regular math class for one period each week (one out of three regular meetings of the class) to meet with two math therapists. The subject of the special class was feelings about math, and the consultants intentionally did not teach any math at all. At first, the teacher was afraid that her class would fall behind because her students were missing one-third of their lessons. Soon she discovered, however, that her students were progressing more rapidly in the two classes they still had with her than they had previously done in three classes per week. Why? Because they were conquering some of the negative feelings that were keeping them from mastering math.

I believe that this talking process is at the heart of the treatment of math anxiety. As we have seen, it helps some people to know that they are not the only ones to suffer from fears of inadequacy about math or science. Moreover, the process of recollection, as stimulated by the math autobiography, can remove old obstacles to learning and provide insight into what is blocking learning now. At first, the participants talk either alone with a counselor or in a group. Later, the talking is transferred to the more private medium of the math diary or the tape recorder, though talking with others may continue. The eventual goal is to understand so well how we get in our own way that we can stop resisting and get down to learning math.

One caveat must be stated. Little is gained by taking people who are worried about not doing well in math and simply making them less worried about it. To tell them only that their answer is a little bit wrong and they should not feel bad about it is not much more than solace. Solace may be necessary, but by itself it won't cure math anxiety. Mathematics mastery is what we're after.

The test of success for any math-anxiety-management scheme will be whether one bites the bullet and signs up for an appropriate course in math, sticks with it, and does well. Since the anxious person on the road to recovery is still well outside his comfort zone, he might better take that course at a nearby community college than at the university, where he would compete with people who are very comfortable with math.

Some community colleges have recently restructured their courses in elementary mathematics to appeal to adults and meet their needs. Arithmetic and algebra are offered as "arithmetic techniques" and "algebraic techniques," and the curriculum advances slowly but surely through "linear mathematics" and "elementary functions" or "probability" to two terms of precalculus. In all, students can profitably spend six semesters getting ready for calculus or statistics. One does not have to try to learn math at the kind of institution or pace at which one might study history or literature. The community college, with its emphasis on teaching and its careful, nonthreatening curriculum, may be just the place for a math-avoidant adult—even someone with a previously earned B.A.—to begin to rebuild her confidence and skills.

Diagnosis and Cure

The first stage in any treatment process is diagnosis. In math therapy, this takes the form of an initial interview, in which the counselor asks about past experiences in math. Out of this questioning comes the mathematics autobiography discussed earlier. How well do you think you understand arithmetic, algebra, and geometry now? What feelings have you been harboring about mathematics? What kind of schooling did you have? Who helped at home? Do you remember what people said to you when you made a mistake in math or when you did unexpectedly well? When did failure begin? Do you feel that you don't know how to read a math textbook? Do you fear that you are slow? When you solve a problem, do you know why? Do you think that you have been away from math for so long that you have forgotten everything?

There are some tests of math anxiety that can be purchased and administered at home, but nothing replaces a well-structured interview.

In the reduction of math anxiety, whether done formally as part of a group or by oneself, diagnosis is a continuing process. Another of the innovative ideas that Bonnie Donady brought to the Wesleyan Math Clinic from her counseling background was the divided-page exercise, sometimes called "double-entry note-taking." Therapists know that it is often "negative self-statements"—things we say to ourselves as we are doing a certain kind of task—that deprive us of competence and confidence in a particular situation. Beginning with the first day of group work, Donady instructed students to keep a record of their feelings (see figure 8-1), particularly the negative ones, on the left-hand side of their notepaper as they worked problems in class or at home.

Under "My feelings/thoughts" they tried to answer questions like "What am I feeling, now, this minute?," "What does this remind me of?," and, later, "What is making this problem difficult for me?" and "What can *I* do to make it easier for myself?" The point of the divided-page exercise, as Donady and many other counselors use it, is to have students learn to *give themselves*

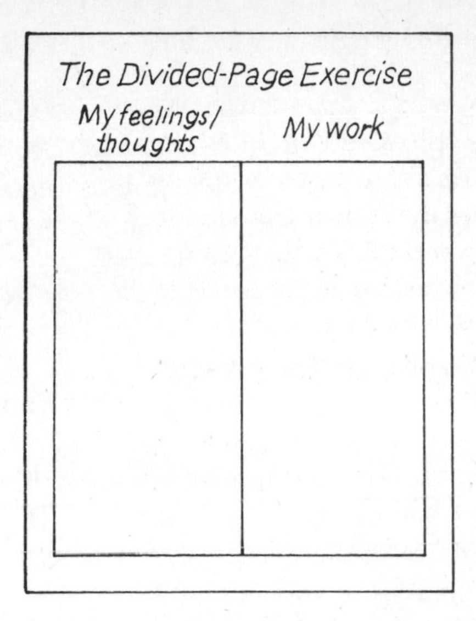

Figure 8-1

permission to explore their own confusion and to find out what is bothering them *specifically* about the problem at hand. "This is just the kind of problem I can never solve" frequently appears on the left-hand side. In discussion, the counselor or that student's classmates might comment that this kind of approach is likely to close more thinking doors than it opens. Or "The numerator is bigger than the denominator. I'm sure my answer is wrong." Not necessarily. As math learners become more and more familiar with their own learning blocks, they become more able to discover self-cures. And having something to do—writing something on the left-hand side of the page—allows them to keep on working.

Perhaps you have already recognized the real purpose of the divided-page exercise. Even when a student is paralyzed, not able to begin to solve the problem, she can still be engaged in thinking about the problem from this other point of view. In a sense, then, the technique forces her to continue working, even after she's "given up." Trying to answer the question "What is making this problem difficult for me?" or "What can I do to make this problem easier for myself?" relaxes the problem's hold on a student and just might give her the time and space to get some idea as to how to solve it.

In time, with increased confidence and understanding of personal hangups, the left-hand side of the page becomes less significant—sometimes even blank. In theory, as long as a student is burdened by negative feelings, she cannot concentrate on the mathematics. And concentration is the key.

"Approaching Math Positively"

Cynthia Arem is an educational psychologist who has teamed up with Debbie Yoklic, a math instructor, to provide math-anxiety reduction for community-college students. Their one-credit course, "Approaching Math Positively," is based on techniques elaborated on in Arem's book, *Conquering Math Anxiety,* a self-help workbook for the math-avoidant.[2] Pima Community College in Tucson labels the course "Math 50," but in fact Arem and Yoklic part company from the more traditional models and do not introduce any new or remedial mathematics during their sessions. Their goal, as attested to by their students—a mix of people concurrently taking any level of math at the college, and students who either dropped out or are too fearful to enroll—is to teach them how to change the beliefs and actions that inhibit their success with mathematics. Stress management, attitude improvement, test-anxiety reduction, visualization techniques, better study skills, and problem-solving *strategies* (but not problem solving itself) are all discussed in this course.

Students talk freely in this class. They are invited to share openly their fears and dislike of mathematics. "I figure God invented calculators, so why do I have to struggle with this?" is a frequently heard lament. "All someone has to do is say the word 'algebra' and I get anxious." For some, anxiety is mixed with anger. "I had a high-school teacher who wouldn't answer my questions. She thought I was dumb in math. I knew I wasn't." For others, there's hope: "I want to like math."

On the second day of class, when students share their math

autobiographies, there are stories of learning and language disabilities that inhibited math understanding: moving from one school to another, interrupting the continuity of education, problems with teachers, problems with parents, even stories of physical abuse in the home that focused on math work. Gathered here, says Jacqueline Raphael, who audited the course for this book, is a curious combination of exceptional students who weren't encouraged to approach math creatively, and some who, for one reason or another, couldn't grasp an important segment of arithmetic, geometry, or algebra and never could catch up.

Building on a technique pioneered by Stanley Kogelman in Mind over Math (the program he ran starting in 1977, with the same name as his book), Arem and Yoklic resist the temptation to explain certain mathematical principles. But they go much further. They try to suggest new approaches, psychological and commonsensical. Yoklic lists math tutors who are particularly able in working with students in lower-level courses. She and Arem suggest courses and course sequences to follow. But, most of all, the two instructors introduce their students to tried-and-true techniques for getting control over their anxiety. "By simply changing the rate and pattern of your breathing during anxiety-provoking situations," Arem says as she walks the students through a breathing exercise, "you can switch from anxiety to calm. You can alter your mind's and body's response to stress and better manage your anxiety."

Arem and Yoklic have a strong bias: They want to breed math winners. Their upbeat, positive, motivating message is a constant theme from first session to last. Some of the students who have taken their course go on to successful careers in engineering and science. A few have even been inspired to become mathematics teachers. But their students are also lucky in the sense that there is a formal math-anxiety-reduction course available in their local community college. Many adults do not have this option. They could try to lobby for such an offering—community colleges, unlike state and private institutions, are supposed to be responsive to locals' needs—but until specialists are found to *teach* math-anxiety reduction, they may have to teach themselves.[3]

On Your Own

If, as some have claimed, mathematics anxiety is caused by nothing more than inadequate math preparation, then one way to overcome it is to get back into math at the appropriate level as soon as possible. Some will be deterred by fear, believing that, since they stopped at, say, the sixth-grade level, it would take them seven to nine years to reach the point where they could tackle the statistics and calculus they need. Not so. With the sophisticated mental equipment we have as adults, sharpened by years of experience in learning, we can make up time extremely quickly.

Three areas suggest themselves as suitable for training and practice: spatial skills, number play, and word-problem solving.

One way to approach spatial skills is simply to learn how to solve those figure problems that are given on tests. Paul Jacobs, a specialist in psychological testing, has written a book with numerous examples of two-dimensional-figure problems and hints on how to solve them.[4] Although this type of exercise might first appear altogether too test-oriented, the fact is that doing enough of these figure problems under the author's supervision does increase skills in spatial visualization. The same is true for certain kinds of spatial games.

The second approach involves using materials that help us visualize mathematical relations. Many of these materials did not exist or were not widely used when I was in school. The Montessori method has long used concrete materials to illustrate the number line and other kinds of numerical and fractional relationships. Although high-school students and adults might incorrectly think of these materials as beneath them, Persis Herold, author of *The Math Teaching Handbook,* recommends them highly for all ages.[5] These materials help make palpable the meaning of a fractional part, an area, even a change in one variable.

One good example is the geoboard, a simple rectangle usually made of wood, dotted with pegs or nails in some regular order (like the stars on the American flag). Using rubber bands, which can be stretched around three or more nails to fashion some kind of polygon, we can create any number of shapes and begin to see quite realistically how area relates to perimeter, and what happens when the perimeter changes. Area seems quite tangible on a geoboard: It is not just some arbitrary manipulation of length times width. It is easy to underestimate the value of such an experience, for the real excitement in mathematics is in the abstractions. But if we actually get our hands on these elements, we are more likely to understand and hence remember what area is all about. Jane S. Stein at Duke University's Continuing Education Program, like Persis Herold, has used geoboards, tangrams, and other "elementary-school materials" quite successfully in a program for adult women.

Dotted paper, sold by some companies but easily made at home, can do almost as well as the geoboard. One can draw lines connecting the dots to form shapes. Graph paper is not obsolete, either. By graphing functions, we can get used to what happens when measurements are compared.

Once upon a time, bisecting an angle or "dropping a perpendicular" in geometry class was an exercise in frustration. I was taught to use a compass and a straight edge. One of the reasons I never appreciated that the lines bisecting the three angles of a triangle meet in the middle at a single point is that my bisections never did meet. I was a messy, inaccurate draftsperson, then as now. But with a newly devised miracle tool, appropriately named Mira, bisecting an angle and transforming a geometric shape are a snap. The Mira is a single plastic surface that is both transparent and reflecting.

Number play might take several forms. Since getting involved again with mathematics, I have found it useful to expand my multiplication table up through the 15s. This makes me feel secure in estimating many 2-digit multiplications. To know that $15 \times 7 = 105$ quite as readily as to know that $6 \times 7 = 42$ gives me a better sense of the size of things. Another good group of numbers to memorize are the most common fractions and their percent approximations: 8 percent is roughly the same as 1/12; 6 percent is roughly the same as 1/16. These come in handy for making rough percentage estimates and for going quickly from percents to fractions and back again.

FRACTION	PERCENT *(approximate in some cases)*
1/2	50 percent
1/3	33.3 percent
1/4	25 percent
1/5	20 percent
1/6	16.7 percent
1/7	14 percent
1/8	12.5 percent
1/9	11 percent
1/10	10 percent
1/11	9 percent
1/12	8.3 percent

It is not always necessary to memorize these numerical equivalents. Like the man who carries a multiplication table around in his wallet, we, too, can create sets of tables, formulas, or even problem-solving hints on cards that we carry around. No one expects us to figure out what day of the week it is from the date and year: We are supposed to consult a calendar. In the same way, we can refer to our private tables. Soon this inventory might even slip quietly into our conscious memory.

We do not know enough about the workings of the brain to say for certain what the best preseason training might be. For some, puzzling through Letter Divisions (see p. 148) may be the best practice. For others, just thinking about numbers and their relationships will help. For still others, reading in the history of mathematics can be least painful. Instructors in math clinics advise everyone to do word problems regularly every day. That's probably not a bad way to begin.

Reading Mathematics

Alan Natapoff says that our math learning is not "oral" or "aural" enough, and attempts to compensate for this in his remedial classes by having people speak out loud what they are being told. Another criticism of conventional math teaching is that we are not taught to read mathematics so that we can learn it on our own. Those of us who do not go on in math, it is true, never really learn to study mathematics the way we study other subjects. Studying math as we do it in the lower grades consists of review and drill. Studying on our own involves learning to understand new material without the help of a teacher. To do this effectively, we have to learn to read math, and that is what many of us do not know how to do.

One way to begin to read mathematics is to recognize how we read other subjects. Most texts in the humanities and even in the social sciences state important ideas and facts more than once. Therefore, those of us who enjoy reading this kind of material learn to read quickly, even to skim, to get the gist of what is intended, and we do not worry too much about missing something. Chances are, if a fact or an interpretive statement is important, it will be repeated or paraphrased. Topic sentences, paragraphs, the structure of a well-written essay are all signposts to tell the rapid reader where to slow down and even to stop. But reading mathematics is reading for immediate mastery. Things are stated only once and must be well understood before we move on. The process is so different from ordinary, even serious reading that the word "reading" might itself be misapplied.

Mathematicians tell me they would not think of tackling a math book without pencil and paper. They try to sketch, if possible, what is being said. They stop and imagine examples that would illustrate the problem at hand. They ask themselves questions and try to answer them. Above all, they move slowly, very slowly, over each part of each statement.

Relapse

One barrier falls and the next one is higher.
—old Chinese proverb

Let's say you're on your way. You have gone through one set of math-anxiety workshops and signed up for a review algebra course. You're doing puzzles and word problems every day and figuring out graphs and tips and percentages. You feel good about math and even better about yourself in general, and then, one day, in class or out, you encounter a problem or procedure or quantitative analysis that is not just unfamiliar; it is impossible! All the old misgivings and "I can't"s may come flooding back again, and it may seem like sudden death.

The fact is, math is not easy, and math anxiety is experienced not only by people like the person you used to be but by experts, too; and it does not occur only at the elementary levels of arithmetic. Able students of math who succeed in high school get the feeling of sudden death doing calculus. Does this mean math anxiety cannot be "cured" or eliminated after all? Probably. But if mathematicians also feel anxiety when confronting a problem of a sort they have never seen before, then the way out is not to deny the anxiety but to manage it. Self-talk, time out for careful consideration of the issue, going for help to someone who has been helpful in the past, can get you over that barrier.

In some cases, the formerly math-anxious will become so attached to a particular teacher, even to a particular text that finally helps them make sense of the material, that they have something akin to withdrawal symptoms when they have to move on to the next class or the next book. Not yet entirely autonomous, the learner attributes her success to some extraneous element, the teacher or the book, instead of to herself. An apparent relapse, then, might be simply an old response to unfamiliar terrain.

It is important to go through this at least once, because managing anxiety is just that: experiencing anxiety and mastering it. Besides, given the inertia in the school system, adults as well as children will probably have to cope with imperfect teachers and texts for a long time to come. Competence only in the ideal, protected situation is not enough. The real goal is autonomy.

Notes

The following math-anxiety programs and materials have been observed and consulted in preparing this chapter:

Algebra I, National University, courtesy of Frances Rosamond.
Math Anxiety Clinic Course, University of Arizona, courtesy of Gary Horne, Pamela Reavis, and Mary Ellen Hunt.
Reentry Math, a program for adult learners in Augusta, Maine, courtesy of Jean Smith.
Math Clinic, University of Utah, courtesy of Kathryn Brooks and Ashley DuLac.
Math 50, Pima Community College, Tucson, Arizona, courtesy of Cynthia Arem and Debbie Yoklie.
Overcoming Math Anxiety, a consulting and training program, in New York City, run by Susan B. Auslander.
Archives of the Wesleyan Math Clinic, courtesy of Bonnie Donady, Jean Smith, Susan Auslander, and Robert Rosenbaum.
Film, *Math Anxiety: We Beat It, So Can You,* 1980, about the Wesleyan Math Clinic, available from Educational Development Corporation, 55 Chapel St., Newton, Mass. 02160.
Book, John Allen Paulos, *Innumeracy: Mathematical Illiteracy and Its Consequences* (New York, Hill and Wang, 1988).
Book, John Allen Paulos, *Beyond Numeracy: Ruminations of a Number Man* (New York: Knopf, 1990).

1. Adapted from Stanley Kogelman and Joseph Warren, *Mind over Math* (New York: Simon and Schuster, 1978).
2. Cynthia Arem, *Conquering Math Anxiety: A Self-Help Workbook* (Pacific Grove, Calif.: Brooks/Cole Publishing, 1993).
3. See appendix for a composite math-anxiety rating scale and for more information about how to order the most common ones.
4. Paul Jacobs, *Up the IQ* (New York: Wyden Books, 1977).
5. Persis Herold, *The Math Teaching Handbook,* available from the author at 3311 Macomb St., Washington D.C. 20008.

9 Afterword

by Sanford Segal

Professor of Mathematics,
University of Rochester

To begin, a personal reminiscence may be appropriate.

In my college generation, some people enjoyed mathematics and were reasonably good at it, and others didn't. Deciding on a mathematical career but enjoying many things besides mathematics, including languages and history, I never thought much about this at the time. While in graduate school, I made friends whose intellect I admired, but who admitted to being baffled by adding or dividing fractions. Multiplication was no problem, because one "followed the rule," which was a rule sufficiently similar to the rule for whole numbers to be "doable" even if there was no need, desire, or emphasis for understanding what was done. A number of these friends were women, who thought (because that is how they had been taught) that manipulation of fractions followed arbitrary and hence scary rules—they didn't understand fractions, and so—indeed, somewhat sensibly—couldn't do fractions. Since some of these were very bright people, I simply thought this an anomaly; in fact, I can remember trying to explain to one such friend *why* addition of fractions made sense, but in vain: she was already convinced it was not worth the effort. Although I had attended an all-male college as an undergraduate, there were always women in my graduate-school classes. I was not concerned with their numbers, and at the time did not know (and would not have cared) that the percentage of American female Ph.D.'s was then in the low single digits. As a young mathematician in the mid-sixties, I would go to social gatherings and be asked by strangers what I "did"; when they learned I taught mathematics, the usual response, and almost invariably from women, was some version of "Oh, you must be very bright!" coupled with "I could never do that in school!" Needless to say, something of a pall fell over the conversation. So much so that I quickly developed the defensive habit of quickly saying, after truthfully acknowledging my profession, "But I'm interested in many other things, also." True, but not entirely helpful. All these phenomena still happen now, some thirty years later (though the percentage of American female Ph.D.'s is now in the mid-twenties), but perhaps with diminishing frequency—the atmosphere is changing thanks to an effort in which Sheila Tobias and the first edition of this book were among the pioneers.

A change began in the mid-seventies. John Ernest, coincidentally a former colleague of mine at the University at Rochester, wrote a pamphlet called *Mathematics and Sex* which was widely distributed by the Ford Foundation. Sheila Tobias wrote an article in *Ms.* entitled "Math Anxiety: Why Is a Smart Girl Like You Counting on Your Fingers?" Why this change, and why then? In part, it was the burgeoning feminist movement waking up to the fact that some knowledge of mathematics was a critical filter for an increasing variety of jobs as our world became increasingly technological and mathematized. In part, also, I believe it was because education, especially but not solely mathematical education, went downhill in the seventies as both secondary and collegiate educators yielded to the irrelevant cry of "relevance" and weakened their curricula while eliminating requirements. Whatever the much-maligned SAT examinations measure, they showed a steep decline in the abilities that had earlier correlated with success in college. Ironically, at the very time when knowledge of some collegiate mathematics was becoming more valued by employers, the standards for mathematics education were in decline. The economic situation played a role, too. Increasing numbers of married women started going to work to support a middle-class life style, as well as for valid reasons of self-fulfillment. And they wanted good jobs, which increasingly seemed to mean some mathematical knowledge, while the schools were, in general, providing less and less of that, except to a select few. The number of people taking advanced placement examinations has continually increased, while the level of general mathematical education has languished. In addition, even that little mathematics was being avoided by women if they could do so, since they had been socialized from early years to think of mathematics as unfeminine.

In 1977, influenced partly by the Math Clinic at Wesleyan University (where I had been an undergraduate some two decades earlier, when it was all-male), I began a clinic for undergraduates who were "anxious" about mathematics at the University of Rochester. At this university, the male/female ratio was (and is) about 60/40, and an extraordinarily large number of entering freshmen express an interest in some sort of scientifically oriented career. Several things soon became apparent. One was the enormous amount of psychological affect invested in mathematics: the young woman for whom mathematics would always be her punitive fourth-grade teacher. Another was that anxiety about mathematics was not just a female problem: I shall never forget one young man whose father was an engineer, as was his elder sister(!), but who felt frustrated about implicit family expectations and being unable to live up to them. Though the clinic was intentionally and primarily for undergraduates, it was open to people in the community. I was surprised when one of these "drop-ins" was a social acquaintance who worked for a social-science research firm and confessed privately his eternal frustration with mathematics, and how he felt that frustration hampered him in his work.

There is no question, however, that math avoidance was and is more prevalent among women than men, though it disables both. It was feminists like Sheila Tobias who, with justified concern about women, made us realize how debilitating it was for all those who suffered from it. Other educators had noted the phenomenon earlier and failed to excite interest. It took the force of being a part of a feminist agenda, and Sheila Tobias' principal role therein, to make us all aware of something we should have been concerned about much earlier, for both women and men.

Of course, part of the earlier social unconcern was that women were the most obvious sufferers. The term "math anxiety" has become enshrined in the literature, but I have come to think that "math avoidance" may be better, not only because it is descriptive of behavior rather than emotion, but also because "anxiety" could be taken to refer to some innate disability. Indeed, I have known people who have unfortunately adopted exactly this "blame-the-victims" approach to the problem. Whether we call it "math anxiety" or "math avoidance," there is no question that it is a product of social training, and of the social training of women in particular.

Many highly intelligent people avoid mathematics, and are ill-served by that avoidance. Indeed, if one imagines mathematics as a subject in which there is only one way to get a correct answer, in which procedures need to be memorized by rote without any reason for them, in which formulas jangle without rhyme, in which competition to get the "right answer" by any means is the epitome of classroom experience, then avoidance may be in fact intelligent! Sheila Tobias' book is addressed to people, especially but not solely women, who have been damaged by this sort of presentation of mathematics, and consequently think they are "dumb." She not only addresses the psychological issues, but demonstrates that all adults of ordinary intelligence have the capacity for mathematical thinking—in fact, as logical deduction, do it all the time. Mathematics is a language, a very concise language. Like all languages, it requires the application of intelligent practice to learn. Furthermore, the general lack of redundancy in mathematics may make it seem somewhat harder to learn than "natural languages." This book shows, however, that the capacity for learning to "speak" mathematics lies with any of its readers.

The problems that stimulated the first edition of this book have begun to be alleviated—in no small measure because of it. However, they are far from solved, especially for women, who were and still are major victims of the math-avoidance syndrome. This new edition has been updated to address the research of the past fifteen years, some that is sympathetic to its theses, but also some that seems to run counter to them. Its clarity, passion, and enthusiasm are undimmed. It should be as beneficial as its earlier incarnation to a new generation of readers, who will learn that they can not only do mathematics, but enjoy it as well!

Appendix

The following is a composite of confidence and anxiety rating scales that have been developed by Dr. Elizabeth Fennema and Dr. Julia Sherman. A full set of these scales is available for $5.00 from the American Psychological Association, 1200 17th St., N.W., Washington D.C. 20036 They are called the "Fennema-Sherman Scales." The full set includes the following:

1. Attitude Toward Success
2. Math Anxiety
3. Confidence in Learning Math
4. Usefulness of Math
5. Perception of Mother's Attitude
6. Perception of Father's Attitude

7. Perception of Teacher's Attitude
8. Affectance Motivation

Composite Math Anxiety Scale*

For each statement, give a number 1–5 which indicates whether you strongly agree (1) or strongly disagree (5).

I usually have been at ease in math classes.
I see mathematics as a subject I will rarely use.
I'm no good in math.
People would think I was some kind of a grind if I got A's in math.
Generally, I have felt secure about attempting mathematics.
I'll need mathematics for my future work.
I'd be happy to get good grades in math.
I don't think I could do advanced mathematics.
It wouldn't bother me at all to take more math courses.
For some reason, even though I study, math seems unusually hard for me.
I will use mathematics in many ways as an adult.
It would make people like me less if I were a really good math student.
My mind goes blank and I am unable to think clearly when working in mathematics.
Knowing mathematics will help me earn a living.
If I got the highest grade in math, I'd prefer no one knew.
Math has been my worst subject.
I think I could handle more difficult mathematics.
Winning a prize in mathematics would make me feel unpleasantly conspicuous.
I'm not the type to do well in math.
Math doesn't scare me at all.

*Adapted from the Fennema-Sherman Mathematics Attitude Scale, available from the Wisconsin Center for Educational Research, University of Wisconsin, Madison WI, 53706.

Further Reading: Introductory Math Books for Adults

Cynthia Arem. *Conquering Math Anxiety: A Self-Help Workbook.* Pacific Grove, Calif.: Brooks/Cole Publishing, 1993.

Laurie Buxton. *Do You Panic About Maths: Coping with Maths Anxiety.* London: Heinemann, 1981.

Carol Gloria Crawford. *Math Without Fear.* New York: Franklin Watts, 1980.

Robert D. Hackworth. *Math Anxiety Reduction.* Clearwater, Fla.: H. and H. Publishing, 1985.

Peter Hilton and Jean Petersen. *Fear No More: An Adult Approach to Mathematics.* Menlo Park, Calif.: Addison-Wesley, 1983.

Stanley Kogelman and Barbara R. Heller. *The Only Math Book You'll Ever Need.* New York: Facts on File, 1986.

C. A. Oxreider. *Your Number's Up.* Reading, Mass.: Addison-Wesley, 1982.

Elizabeth Ruedy and Sue Nirenberg. *Where Do I Put the Decimal Point?* New York: Henry Holt, 1985.

Britte Immergut and Jean Smith. *Arithmetic and Algebra for Math Anxious Adults.* New York: McGraw-Hill, 1993.

Sheila Tobias. *Succeed with Math: Every Student's Guide to Conquering Math Anxiety.* New York: College Board, 1987.

A particularly good strategy for math-anxious adults is to revisit mathematics as a family activity. A book prepared in Spanish and in English by the Equals Project of the Lawrence Hall of Science ably meets this need: Jean Kerr Stenmark, Virginia Thompson, and Ruth Cossey. *Family Math* (also in Spanish, as *Mathematica para la Familia*). Berkeley, Calif.: Lawrence Hall of Science, 1986.

For teachers, or for those working in a group, there are the "Project Mathematics!" computer-animated mathematics videotapes, available For $11.25 plus postage and $1.50 for extra workbooks, from Caltech Bookstore, Mail Code 1-51, Pasadena CA 91125. The project director is Tom Apostol.